Diss. ETH 13292

Development and Characterization of Glassy Carbon Electrodes for a Bipolar Electrochemical Double Layer Capacitor

A dissertation submitted to the

SWISS FEDERAL INSTITUTE OF TECHNOLOGY
ZÜRICH

for the degree of
Doctor of Natural Sciences

presented by

Artur Braun
Diplom-Physiker (RWTH Aachen)
born October 11, 1965
in Schleiden/Eifel (Germany)

accepted on the recommendation of

Prof. Dr. A. Wokaun, examiner
Dr. O. Dossenbach, co-examiner
Dr. R. Kötz, co-examiner

1999

Autor:

Artur Braun

Labor für Electrochemie

Departement Allgemeine Energieforschung

Paul Scherrer Institut

CH - 5232 Villigen PSI

Schweiz

ISBN: 9781099831812

Bibliographische Information bei der Deutschen Nationalbibliothek

Diese Dissertation ist als Hochschulschrift bei der Deutschen Nationalbibliothek unter den Verbindungen http://d-nb.info/1117595579 und http://d-nb.info/1187462942 hinterlegt.

Bibliographische Information bei der Eidgenössischen Technischen Hochschule Zürich

Diese Dissertation ist bei der ETH Zürich als Doctoral Thesis (Diss. Naturwissenschaften ETH Zürich, Nr. 13292, 1999, Examiners: Dossenbach, Otmar; Kötz, Rüdiger; Wokaun, Alexander) erfasst unter den Verbindungen https://doi.org/10.3929/ethz-a-003840479 und https://www.research-collection.ethz.ch/handle/20.500.11850/144375

Bibliographische Information zu früherer Veröffentlichung

Diese Dissertation wurde von 2002 bis zum 10. Dezember 2019 unter mehreren ISBN-Nummern vermarktet und veröffentlicht als Taschenbuch durch den Verlag: Diplomarbeiten Agentur diplom.de.

Copyright © 2019 Dr. Artur Braun, Zürich, Switzerland

Abbildung auf der Rückseite (Superkondensatoren) gefertigt von Dr. Martin Bärtsch, Paul Scherrer Institut.

Einträge im Katalog der Deutschen Nationalbibliothek

Link zu diesem Datensatz	http://d-nb.info/1117595579
Art des Inhalts	Hochschulschrift
Titel	Development and Characterization of Glassy Carbon Electrodes for a Bipolar Electrochemical Double Layer Capacitor / Artur Braun
Person(en)	Braun, Artur (Verfasser)
Ausgabe	1. Auflage
Verlag	Hamburg : Diplom.de
Zeitliche Einordnung	Erscheinungsdatum: 2002
Umfang/Format	Online-Ressource (pdf)
Andere Ausgabe(n)	Elektronische Reproduktion: ISBN: 9783832454401
Hochschulschrift	Dissertation, Swiss Federal Institute of Technology Zurich, 1999
Persistent Identifier	URN: urn:nbn:de:101:1-201611013609l
URL	http://www.diplom.de/e-book/220995/development-and-characterization-of-glassy-carbon-electrodes-for-a-bipolar (Verlag)
ISBN/Einband/Preis	978-3-8386-5440-9
EAN	9783838654409
Sprache(n)	Englisch (eng)
Anmerkungen	Lizenzpflichtig. - Vom Verlag als Druckwerk on demand und/oder als E-Book angeboten
Sachgruppe(n)	530 Physik

Link zu diesem Datensatz	http://d-nb.info/1187462942
Titel	Development and Characterization of Glassy Carbon Electrodes for a Bipolar Electrochemical Double Layer Capacitor / Artur Braun
Person(en)	Braun, Artur (Verfasser)
Ausgabe	1. Auflage, digitale Originalausgabe
Verlag	Hamburg : Diplom.de
Zeitliche Einordnung	Erscheinungsdatum: 2002
Umfang/Format	Online-Ressource, 256 Seiten (pdf)
Andere Ausgabe(n)	Erscheint auch als Druck-Ausgabe: ISBN: 978-3-8386-5440-9
Hochschulschrift	Dissertation, Swiss Federal Institute of Technology Zurich, 1999
Persistent Identifier	URN: urn:nbn:de:101:1-2019053010362069184879
URL	https://www.diplom.de/document/220995 (Verlag)
ISBN/Einband/Preis	978-3-8324-5440-1
EAN	9783832454401
Sprache(n)	Englisch (eng)
Anmerkungen	Vom Verlag als Druckwerk on demand und/oder als E-Book angeboten
Sachgruppe(n)	540 Chemie

Zusammenfassung

Die thermische Gasphasen-Oxidation (Aktivierung) von Glaskohlenstoff (GK) sowie bipolare monolithische Elektroden aus derart aktiviertem GK wurden untersucht.
Die Untersuchung hatte zum Ziel, den Aktivierungsprozess zu verstehen und Korrelationen zwischen Prozessparametern und Materialgrössen sowie den Elektrodeneigenschaften zu finden. Mit diesem Wissen sollte es möglich sein, die Elektrodeneigenschaften zu optimieren.

Die thermische Aktivierung von GK erzeugt auf seiner Oberfläche einen porösen Film, der mit zunehmender Aktivierung in den GK hineinwächst.
Die Filmdicke wurde an unterschiedlichen GK-Platten und -Folien mit dem Raster-Elektronenmikroskop bestimmt.
Die Mikro- und Mesostruktur der Proben wurde mit Röntgenbeugung und mit Röntgenkleinwinkelstreuung untersucht.
Die innere Oberfläche von aktivierten GK-Proben wurde mit Gasadsorptionsmessungen und mit Röntgenkleinwinkelstreuung bestimmt.
Die Doppelschichtkapazität und der Diffusionswiderstand von GK-Proben wurden mittels Elektrochemischer Impedanzspektroskopie bestimmt.

Untersuchungen an GK mit unterschiedlicher Pyrolysetemperatur zeigen eine Korrelation von Pyrolysetemperatur und maximal erzielbarer Filmdicke: Je höher die Pyrolysetemperatur, umso kleiner die asymptotische Filmdicke.
Röntgenbeugung an GK mit unterschiedlicher Pyrolysetemperatur zeigt eine Korrelation der Pyrolysetemperatur mit Intensität und Halbwertsbreite prominenter Beugungspeaks: Je höher die Pyrolysetemperatur, umso geringer ist die Defektdichte in GK.
Daraus folgt die Korrelation von asymptotischer Filmdicke und Defektdichte, die ein qualitatives Mass für den experimentell schwer zugänglichen, für die Filmbildung aber sehr wichtigen effektiven Diffusionskoeffizienten D_{eff} ist.

Ergebnisse aus der Röntgenkleinwinkelstreuung zeigen, dass die Durchmesser der Mikroporen von GK bei 10 bis 20 Å liegen und dass die Porengrössenverteilung nicht sehr scharf ist.
Ein nicht zu vernachlässigender Teil der Poren liegt daher im Grössenbereich unterhalb 7 Å bis 9 Å, der nicht benetzt werden kann und damit nicht zur Doppelschichtkapazität beiträgt.
Der Diffusionswiderstand hat ein Minimum (etwa 50 mΩcm^2) nach einer bestimmten Aktivierzeit (etwa 30 bis 60 Minuten). Für längere Aktivierzeiten steigt der Diffusionswiderstand monoton.

Infolge der Aktivierung kommt es zu einer Verschiebung der Porengrössenverteilung zu grösseren Durchmessern, wodurch ein weiterer Teil der Poren für Elektrolyte zugänglich wird und die volumetrische Doppelschichtkapazität zunimmt. Impedanzmessungen zusammen mit der Filmdickenbestimmung belegen, dass die volumetrische Doppelschichtkapazität mit der Aktivierdauer auf etwa 100 F/cm^3 zunimmt.

Während der Aktivierung verschiebt sich der (002)-Röntgenreflex zu grösseren 2Θ Winkeln, was auf eine Verdichtung der Graphit-Skelettstruktur des GC hindeutet. Weiter wurde gefunden, dass die innere Oberfläche des Aktivfilms von GK mit zunehmender Aktivierzeit abnimmt, und zwar vermutlich infolge eines Zusammenwachsens kleinerer Poren zu grösseren Poren. Diese Ergebnisse stehen voll im Einklang mit den Aussagen des *falling cards model for hard carbons* nach W. Xing et alii.

Das Potenzgesetz für den Abfall der Intensität der Röntgenkleinwinkelstreuung für grosse Streuvektoren Q hat einen Exponenten nahe bei 3.0 (nicht aktivierter GK: ≈2.5), was darauf schliessen lässt, dass die offene innere Porenoberfläche sowie das Netzwerk des GK-Materials über das gesamte Probenvolumen ausgedehnt sind.

Abweichungen von allen oben genannten Ergebnissen wurden jedoch für GK gefunden, der nur schwach aktiviert ist (kurze Aktivierung bei niedrigen Temperaturen oder geringen Konzentrationen des Oxidationsmittels).
(i) Entwicklung der Kapazität (und damit wahrscheinlich der Filmdicke) folgt einem Exponentialgesetz, (ii) Dichte der Graphit-Skelettstruktur nimmt ab, (iii) Potenzgesetz der Kleinwinkelstreuung tendiert zu einem Abfall mit Exponent 2.0, (iv) Diffusionswiderstand übersteigt den minimalen Widerstand um mehrere Grössenordnungen.
Dieser aktivierte GK stellt offenbar einen Übergangszustand von oxidiertem GK dar.

Dem diffusionskontrollierten Filmwachstum ist ein reaktionskontrollierter Abbrandprozess überlagert, der die mathematische Beschreibung des Systems erschwert.
Es wurde ein Modell für die Aktivierung aufgestellt und eine exakte analytische Lösung für das Filmwachstum bei thermischer Aktivierung von GK Proben mit planarer Geometrie (monolithische Elektroden) und mit sphärischer Geometrie (Kugeln für Pulverelektroden) gefunden.
Die Filmdicke für planare Proben wird durch eine sogenannte *Verallgemeinerte LambertW-Funktion* $\mathcal{G}$ beschrieben. Für lange Aktivierungszeiten (asymptotisches Verhalten) ist die Filmdicke eine Konstante, die im wesentlichen aus dem

Verhältnis des effektiven Diffusionskoeffizienten $D_{eff.}$ von GK und seiner Reaktionsrate k gegeben ist. Die Zeitabhängigkeit ist durch $\mathcal{G}(t)$ gegeben. Die Konzentrationserhöhung des Oxidans beschleunigt zwar die Filmbildung, hat aber keinen Einfluss auf die asymptotische Filmdicke. Die Funktion $\mathcal{G}(t)$ wurde an experimentelle Daten angepasst. Experimentelle Daten und Modell stimmen semiquantitativ überein.

Das Modell und die Lösungen sind universell und lassen sich auf andere Systeme mit reaktionskontrollierten Dimensionsänderungen der Proben und diffusionskontrolliertem Filmwachstum anwenden.

Abstract

The thermal gasphase oxidation (activation) of Glassy Carbon (GC) and bipolar monolithic electrodes as prepared from thermally activated GC were investigated. The goal of these investigations was to understand the thermal activation process and to find a correlation between process parameters plus GC materials properties and the electrode performance. With this knowledge it should be possible to optimize the performance of GC electrodes.

Thermal activation creates a porous film on the surface of GC. With proceeding activation the film grows into the GC.
The film thickness of various GC plates and sheets was determined with scanning electron microscopy.
The microstructure and mesostructure of the samples were investigated with X-ray diffraction and Small Angle X-ray scattering.
The internal surface area of activated GC samples were determined with gas adsorption measurements and Small Angle X-ray scattering.
The electrochemical double layer capacitance and the diffusion resistance of GC samples were determined with electrochemical impedance spectroscopy.

Studies on GC with different temperature of pyrolysis reveal a correlation between pyrolysis temperature and asymptotic active film thickness: The higher the pyrolysis temperature, the smaller the asymptotic film thickness.
X-ray diffraction on GC with various pyrolysis temperatures exhibits a correlation between temperature of pyrolysis as well as intensity and full width at half maximum of prominent diffraction peaks: The higher the pyrolysis temperature, the smaller the defect density in GC.
From the two latter findings, a correlation between asymptotic film thickness and defect density is found. The defect density is a qualitative, macroscopic and global measure for the effective diffusion coefficient $D_{eff.}$, which is an essential for the film growth, but experimentally difficult to measure.

Results from the small angle X-ray scattering reveal that the micropores in GC have a diameter of about 1 Å to 2 Å and that the distribution of the diameter is not very sharp.
Considerable part of the micropores has diameters smaller than 7 Å to 9 Å, which cannot be wetted by electrolytes and therefore do not contribute to the capacitance.
The diffusive resistance has a minimum (at about 50 mΩcm^2) after a specific activation time (around 30 to 60 minutes). For larger activation times the resistance increases monotonously.

Due to the activation the pore size distribution is shifted towards somewhat larger diameters, with the result that an additional part of the previously smaller pores becomes accessible for the electrolyte and the capacitance increases. Impedance measurements in combination with film thickness determination prove that the volumetric capacitance increases during activation towards a value of around 100 F/cm^3. During activation the X-ray diffraction (002)-peak shifts towards larger 2Θ angles, which indicates a densification of the graphite skeleton structure of GC. As a further result it was found that the internal surface area of the active film decreases upon activation, probably because of growth and coalescence of pores on the cost of smaller pores. The two latter results are fully in line with the recently proposed *falling cards model for hard carbons* by W. Xing et alii.

The exponent of decay of the scattered small angle X-ray intensity for large Q values was found to be very close to 3.0 for fully activated GC (non-activated GC: $\approx$2.5), which indicates that the internal surface area and the GC are extended through the whole sample volume.

Deviations from all findings above were found for GC activated only weakly (short oxidation at low oxidant concentration or low oxidation temperature):
(i) Evolution of capacitance (and therefore probably also film thickness) follows an exponential law, (ii) graphite skeleton density decreases, (iii) exponent of decay tends towards a minimum of 2.0, (iv) diffusion resistance exceeds the minimum resistance by several orders of magnitude.
The weakly activated GC obviously represents a transition state of oxidized GC.

During activation the overall sample experiences a burn-off, which is controlled by chemical reaction, and a film growth, which is controlled by diffusion of reactants. The superposition of both processes complicates the mathematical formulation of the system considerably.
A mathematical model for the thermal activation of GC was proposed and an exact analytical solution for the evolution of film thickness on GC with plane geometry (monolithic electrodes) and spherical geometry (spheres for powder electrodes) was found.
The film thickness on plane electrodes is described by a so-called *Generalized LambertW-Function* $\mathcal{G}$. For large activation times (asymptotic behaviour) the film thickness is a constant, which is in given by the ratio of the effective diffusion coefficient D_{eff} of GC and its reaction rate k.
The time dependence of the film thickness evolution is expressed by $\mathcal{G}(t)$. An increase of the oxidant concentration accelerates the film growth, but the asymptotic film thickness is not affected by the concentration. The function $\mathcal{G}(t)$ was fitted to experimental data. Experimental data and the model exhibit reasonable agree-

ment.

The model and its solutions are universal and may be extended and also applied to other systems having reaction controlled changes of sample dimensions and diffusion controlled film growth.

Contents

1 Introduction   23

2 Electrochemical Double Layer Capacitors   27
    2.1 Classification of Electrical and Electrochemical Power Sources . . 27
    2.2 Fundamentals 29
       2.2.1 The Electrochemical Double Layer 29
       2.2.2 Working Principle of Supercapacitors 30
       2.2.3 Design of EDLC with Bipolar Electrodes 33
    2.3 Applications for Supercapacitors 35
    2.4 Capacitor World Market 36

3 Glassy Carbon   39
    3.1 On Activated Glassy Carbon 42
       3.1.1 Thermal Activation of Glassy Carbon 45

4 Experimental   47
    4.1 Sample Preparation 47
    4.2 Thickness Determination 48
       4.2.1 Determination of the Sample Thickness 48
       4.2.2 Determination of the Film Thickness 49
    4.3 X-ray Diffraction 50
    4.4 Small Angle X-ray and Neutron Scattering 52
       4.4.1 Small Angle X-ray Scattering 52
       4.4.2 Small-Angle Neutron Scattering 55
    4.5 Electrochemical Characterization 55
       4.5.1 Cell Setup 55
    4.6 Nitrogen Gas Adsorption 56

5 Data Analysis   57
    5.1 X-ray Diffraction 57
    5.2 Small Angle Scattering 62

	5.3	Electrochemical Characterization	74
		5.3.1 Cyclic Voltametry	74
		5.3.2 Electrochemical Impedance Spectroscpy	76
		5.3.3 Diffusive Resistance of Porous Electrodes	82

6 Results and Discussions 85

- 6.1 Thickness and Mass Changes during Activation 85
  - 6.1.1 Weight Loss and Thickness Changes 87
  - 6.1.2 Film Thickness 100
- 6.2 A Model for the Film Growth 101
  - 6.2.1 Introduction 103
  - 6.2.2 Establishing the Model 104
  - 6.2.3 Experimental 111
  - 6.2.4 Results and Discussion 111
  - 6.2.5 Conclusions 117
- 6.3 Influence of Activation Parameters on Electrode Performance . . . 119
  - 6.3.1 Influence of Reaction Time and Temperature 119
  - 6.3.2 Influence of O_2-Concentration 141
  - 6.3.3 Capacitance after Weak Activation 142
  - 6.3.4 Capacitance of Thermally Activated GC with Different HTT 145
  - 6.3.5 Capacitance of Thermally Activated GC from Capton® Foils 159
  - 6.3.6 Impact of Reduction on Electrode Performance 161
  - 6.3.7 Estimation of Capacitance by Geometrical Considerations 164
  - 6.3.8 Influence of Electrolyte Temperature on Capacitance and Resistance 166
- 6.4 Comparison with Electrochemically Activated Glassy Carbon . . . 171
  - 6.4.1 General Remarks 171
  - 6.4.2 Electrochemically Treated Thin GC Sheets 173
- 6.5 Structure and Structural Changes in GC 182
  - 6.5.1 Structural Differences in GC 182
  - 6.5.2 Structural Changes During Activation 189
- 6.6 Small Angle X-ray Scattering on Glassy Carbon 194
  - 6.6.1 SAXS on Various Non-Activated GC 194
  - 6.6.2 SAXS on Thermally Activated GC 200
  - 6.6.3 Consideration of Plausibility 213

7 Conclusions 217

A	**Activation of Glassy Carbon Powder**	**221**
	A.1 Experimental	221
	A.1.1 Oxidation of GC Powder	222
	A.2 Evolution of Surface Area during Activation	223
	A.2.1 Introduction of the Model	224
	A.2.2 Results and Discussion	227
B	**Film Growth Model for Spherical Particles**	**237**

List of Figures

1.1	Prototype Supercapacitors developed at PSI.	25
2.1	Schematic Ragone diagram.	28
2.2	Electrode, electrolyte and double layer.	31
2.3	Sketch of potential at electrode, double layer and electrolyte.	32
2.4	Sketch of a capacitor stack with bipolar electrodes.	34
2.5	Capacitor market share in the last decade.	36
3.1	Escher's Symmetry Drawing No. 133.	39
3.2	Schematic sketch of low HTT GC microstructure.	41
3.3	Schematic sketch of high HTT GC microstructure.	41
4.1	Diffraction peaks of Carbon Reference.	51
4.2	Diffraction peaks of Carbon Reference.	51
4.3	Bulk material correction in SAXS	54
4.4	Electrochemical cell setup.	56
5.1	R-ratio of G-type GC.	58
5.2	R-ratio of K-type GC.	58
5.3	2-dimensional SAXS patterns of GC.	63
5.4	Guinier-plot of 1 mm K-type and G-type GC.	64
5.5	Porod-plot of scattering curves.	71
5.6	Potential/time relationship at a working electrode during cyclic voltammetry.	74
5.7	Cyclic voltamogram of an activated SIGRADUR®K sample.	75
5.8	Schematic of Impedance.	77
5.9	Representation of the electrochemical interface.	79
5.10	Series circuit of a resistance R_s and a capacitance C, representing the most simple capacitor electrode.	80
5.11	Representation of the electric circuit of an EDLC.	81
5.12	Determination of diffusive resistance of porous electrodes.	83
6.1	SEM micrograph and schematic of a partially activated GC sample.	86

6.2	Decrease of sample mass during activation.	88
6.3	Thickness change of K 1mm GC samples as a function of reaction temperature.	90
6.4	Oxidation rate of various carbons in Arrhenius plot.	91
6.5	Weight increase due to humidity and gas adsorption in activated GC.	93
6.6	Mass decrease and curve fit.	95
6.7	Schematic of mass density distribution in activated GC sheets.	97
6.8	Sample surface morphology after activation.	99
6.9	Plot of complex branches of the LambertW function.	108
6.10	Measured and calculated film thicknesses in SIGRADUR®K with 60 microns nominal thickness.	113
6.11	Measured and calculated film thicknesses in SIGRADUR®K with 1 mm nominal thickness.	113
6.12	Film thicknesses of SIGRADUR®K and GC K800 with 100 μm thickness.	114
6.13	X-ray diffractograms of SIGRADUR®K with 1000 μm and 55 μm thickness.	118
6.14	Capacitance of activated SIGRADUR®K with 60 microns nominal thickness.	120
6.15	Volumetric capacitance of SIGRADUR®K with 60 μm thickness.	121
6.16	Capacitance of SIGRADUR®K with 1 mm thickness.	122
6.17	Capacitance of SIGRADUR®K with 1mm thickness for various activation temperatures.	124
6.18	Impedance spectra in Nyquist-representation.	128
6.19	Diffusive resistance of SIGRADUR®K with 60 microns thickness.	128
6.20	Diffusive resistance of SIGRADUR®K with 100 microns thickness.	129
6.21	Diffusive resistance of SIGRADUR®K with 1000 microns thickness.	129
6.22	Diffusive resistance of Capton K 150 GC.	131
6.23	Capacitance frequency response of SIGRADUR®K activated at various activation temperatures.	134
6.24	Frequency response of GC samples activated at 400 °C and 450 °C.	136
6.25	Frequency response of GC samples activated at 500 °C and 550 °C.	138
6.26	Capacitance of K800 GC 1mm, oxidized at 400°C.	139
6.27	Frequency response of the capacitance of K800 GC 1mm, oxidized at 450 and 500°C.	140
6.28	Capacitance depending on gas concentration.	141
6.29	Capacitance of GC K800 and SIGRADUR®K, 1 mm thickness.	146
6.30	Pyrolysis schedule.	152

LIST OF FIGURES

6.31 X-ray diffractograms of samples with $110\mu m$ thickness with HTT 800°C and 1250°C. 153
6.32 X-ray diffractograms of samples with 1 mm thickness with HTT 800°C, 1000°C and 1250°C. 154
6.33 Frequency response of capacitance of SIGRADUR®K and K1250, activated at 450°C. 155
6.34 Frequency response of capacitance of SIGRADUR®K and K1250, activated at 500°C. 155
6.35 Plot of the SOAT temperature for best activation depending on pyrolysis temperature. 157
6.36 Capacitance of activated regular K100 GC and Capton K150 GC. . 159
6.37 Frequency dispersion of the capacitance of Capton K150 GC. . . . 160
6.38 Reduction current as a function of reduction time. 162
6.39 Capacitance and resistance as a function of electrolyte temperature. 166
6.40 Specific resistance of 3 molar sulfuric acid depending on temperature. 168
6.41 Comparison of diffusive resistance and electrolyte resistance. . . . 168
6.42 Comparison of electrolyte resistance and R_{Diff}. 170
6.43 Cyclic voltamograms of electrochemically treated SIGRADUR®G. 175
6.44 SANS Porod plots of electrochemically treated SIGRADUR®G, activated 1 hour at 2.07 Volt. 176
6.45 SAXS curves of electrochemically treated GC G 60 microns. . . . 177
6.46 SAXS Guinier plot of electrochemically treated thin SIGRADUR®G sheets. 180
6.47 Diffractograms of GC with 1 mm thickness and different pyrolysis temperature. 182
6.48 R-ratio for various GC samples with 1 mm thickness and different HTT. 183
6.49 FWHM of (002) peak in GC as a function of pyrolysis temperature. 184
6.50 Crystallite size L_c vs. HTT 185
6.51 X-ray diffractograms of GC with 1000, 100 and 55 μm thickness. 185
6.52 X-ray diffractogram of GC type G with 46 μm thickness. 188
6.53 Interlayer distance of GC as a function of HTT. 188
6.54 XRD intensity as a function of sample thickness. 189
6.55 Shift of (002) peak in GC K 800 with sample thickness. 190
6.56 Shift of (002)-peak in GC K 1000 during oxidation. 191
6.57 Guinier-plot of 1 mm SIGRADUR®K and SIGRADUR®G. . . . 194
6.58 SAXS curve and fitted curve. 196
6.59 Lognormal distribution of pore sizes of various GC samples. . . . 196
6.60 Micropore radius distribution of K800 GC with different thicknesses. 199

6.61	Pore radii distribution of various GC samples.	199
6.62	Log-log plot of thermally activated 1 mm SIGRADUR®K.	201
6.63	Log-log plot of thermally activated 1 mm SIGRADUR®K.	202
6.64	Log-log plot of SAXS curves for activated SIGRADUR®K with 60 microns thickness.	204
6.65	Change of micropore radius distribution during activation	206
6.66	Porosity of SIGRADUR®K with 60 microns thickness vs. activation time.	208
6.67	Evolution of internal surface area during activation.	209
6.68	Comparison of X-ray density and exponent of decay.	214
6.69	Schematic for an arrangement of voids in GC.	216
A.1	Distribution of GC powder SIGRADUR®.	222
A.2	Distribution of GC powder SIGRADUR®K.	223
A.3	Distribution of GC powder SIGRADUR®.	224
A.4	Illustration of changes in particle radii.	225
A.5	X-ray diffractogramms of GC powder.	228
A.6	X-ray diffractogramms of GC plates of 1 mm thickness.	229
A.7	Burn-off of SIGRADUR®K powder	231
A.8	Measured and calculated BET surface area of SIGRADUR®K powder.	232
A.9	Particle size distribution and bimodal fitted curve for SIGRADUR®G powder.	234
A.10	Measured and calculated BET surface area of SIGRADUR®G powder.	235
B.1	Schematic representation of film growth in GC spheres.	238
B.2	Phase portraits of the ODE for spheres for various diffusion coefficients D.	240
B.3	Evolution of film growth, sample shrinking and unreacted core shrinking in a reacting sphere.	242
B.4	Plotted film thickness curves as a function of reaction time.	247

List of Tables

5.1 Variation of crystal parameters with HTT. 60
5.2 Geometry factors for the conversion of R_g into the actual radius R. 66
5.3 Asymptotic behaviour of various scatterers. 70
5.4 Exponent of decay for various objects. 73
5.5 Resistivities of various electrotechnical elements. 78
6.1 Best fit values for film growth 112
6.2 Capacitance of SIGRADUR®K. 125
6.3 Diffusion resistance of SIGRADUR®K. 132
6.4 Bending Point of SIGRADUR®K activated at 400°C and 450°C. . 135
6.5 Capacitance of SIGRADUR®K vor various gas concentrations. . . 142
6.6 Capacitance of SIGRADUR®K for various gas concentrations. . . 144
6.7 Capacitance of thermally activated SIGRADUR®G. 148
6.8 Capacitance and diffusive resistance of thermochemically oxidized SIGRADUR®G with removed skin. 149
6.9 Capacitance of thermochemically oxidized GC SIGRADUR® G, oxidized in oxygen for 1 hour. 150
6.10 Active film growth rates for electrochemical activation. 172
6.11 Oxygen content on SIGRADUR®K after thermal activation. . . . 174
6.12 Oxygen content on GC after electrochemical treatment. 174
6.13 Sample thickness and mass per sample area. 178
6.14 Diffusive resistance of electrochemically treated SIGRADUR®G sheets. 179
6.15 XRD data for various non-activated GC. 187
6.16 Micropore gyration radius and mass density of unactivated GC. . 195
6.17 Micropore fitting results. 205
6.18 Comparison of mean values for micropore radius, surface area and micropore volume. 211
6.19 Exponent n of decay for scattering curves of 60 μ SIGRADUR®K. 213
A.1 Burn-off rates and active film growth rates for various GC sheets of different thickness 230

A.2　Fitting parameters for SIGRADUR®G powder. 233

Chapter 1

Introduction

The scope of this thesis was to study the thermochemical gas phase activation of Glassy Carbon (GC) for the utilization as an electrode material for Electrochemical Double Layer Capacitors (EDLC).
Effort should be made to yield a detailed understanding of the activation process and the GC structure and to find correlations of process parameters and electrode performance.
In particular,
(i) electrodes as obtained by thermal gas phase oxidation of GC were prepared and characterized, and
(ii) the oxidation process itself were studied.
A detailed study of materials and processes is presented in this work.

The characterization of the material was carried out with classical tools such as X-ray Diffraction, Impedance Spectroscopy, Gas Adsorption, but also with modern large scale facilities providing synchrotron radiation and neutrons.
It is shown that the structural properties of the Glassy Carbon affect the electrode performance by far.
A quantitative model and process description is presented which in principal allows to taylor electrodes with specific properties.

The topic of this thesis arose from a project funded by the Swiss Federal Institute of Technology within the Swiss Priority Program on Materials Research.
The goal of this project was to develop an EDLC for the utilization in locomotive trains developed and produced by the company Asea Brown Boveri (ABB).
The advantage of such novel EDLC would be that the weight and volume of a locomotive which still uses conventional capacitor benches decreases.
The benefit of a *light* locomotive train is not so much related to the energy consumption of a public means of transportation. The real advantage would be that

the overall infrastructure of the train and also the track could be constructed and manufactured more cheaply. Sofar, the heavy weight of the locomotive requires expensive materials (high quality steels) and design concepts [1].

Glassy Carbon (GC) is an electrode material very well suited for electrochemical double layer capacitors (EDLC) [1] [2].
GC is electronically conducting [2], has a small specific weight, is resistive against corrosion and may be manufactured in plates, thin sheets and as powder.
Additionally, it may be produced in different shapes such as bottles and cylinders. GC is highly porous with closed pores in nanometer size and impermeable to gases. By an appropriate activation process (usually an oxidation: chemical, wetchemical, electrochemical, thermochemical) the closed porosity can be converted into an open porous structure.
When the GC with open pores is immersed into an electrolyte, an electrochemical double layer is created at the interface between the walls of the open pores and the electrolyte. By polarization, electric energy can be stored in this double layer.

The high porosity and large internal surface area and good electronic conductivity of an activated GC sample in principal could meet the requirements for an application as an electrode of an EDLC with a high energy density and power density.
The two plane areas of a GC sheet or disk can be activated in a single production step, and the resulting sheet can be regarded as a monolithic bipolar plate assembly.
A prerequisite for an EDLC is an as low as possible equivalent series resistance (ESR). The internal surface area of the GC has to be opened by activation in such a way that the electronic and ionic conductivities do not decrease.
With a detailed understanding of the processes taking place on microscopic and mesoscopic scale during activation it was deemed possible to optimize the activation process with respect to the required properties of the electrode.
Capacitance, ESR, frequency response of the capacitance, energy density and power density of the EDLC are determined by the thickness, pore structure, surface properties and conductivity of the active layer of the GC.
During the development and search for optimum activation parameters it has to be kept in mind that the manufacturing of the electrodes is going to be carried out in a large scale production with economical and ergonomic efficiency and ecological responsibility.

---

[1] Electrochemical Double Layer Capacitors are frequently also called Supercapacitors, Ultracapacitors, Powercapacitors and Boostcaps. All these names are registered trademarks and brands of companies such as Matsushita, Panasonic, Maxwell Industries and Montena S.A.

[2] Correct speaking it is a narrow gap semiconductor.

This research is embedded in the project *Supercapacitors* with funding from the Swiss Priority Program on Materials Research (1996 - 1998) and the Swiss Commission for Innovation and Technology (1999 - 2001). In December 1998 the

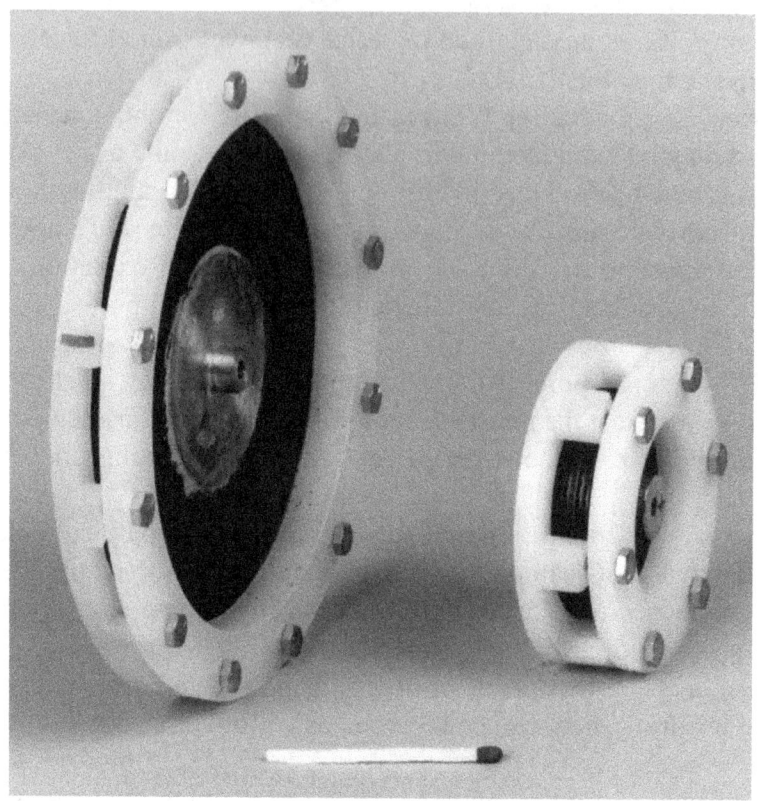

Figure 1.1: 2 prototype supercapacitors developed at PSI. On the right side a 5 Volt capacitor stack with thermally oxidized glassy carbon electrodes with 20 cm^2 effective electrode area. Benchmark data were 0.45 Farad, 26 mΩ ESR, 38 kW/liter power density and 0.6 kJ/liter energy density. The stack was cycled 10^5 times during 1110 hours of maintenance between 0.0 and 5 Volt. On the left side a 1 Volt capacitor with scaled up GC electrodes of 12 cm diameter. Photography by courtesy of Dr. Martin Bärtsch, PSI.

state-of-the-art prototype supercapacitors developed at PSI exhibited a power density of 38 kW/liter - the highest power density for an electrochemical double layer capacitor ever reported [3].

Chapter 2

Electrochemical Double Layer Capacitors

2.1 Classification of Electrical and Electrochemical Power Sources

Power sources [1] are classified by their energy density (Joule/liter) and power density (Watt/liter).

Batteries can store a high amount of chemical energy, which is released as electrical energy upon usage due to a chemical (faradaic) reaction.

Capacitors can store quite less energy in comparison, but they can store and release the energy faster than batteries.

The power density is the ratio of the energy density and the time constant τ. The time constant of a capacitor is the product of its capacitance and resistance. Typically, batteries have smaller power densities than capacitors.

However, due to the different nature of energy storage mechanisms in batteries (chemically) and capacitors (electrostatically), the capacitors have a smaller energy density. This relation is displayed in a so called *Ragone Diagram*, where the power densities of power sources are plotted versus their energy densities. Therefore, the region of high energy density and small power density is covered by the batteries, whereas the capacitors cover the region of small energy and high power density.

There is a gap in the Ragone diagram between these two types of power sources, which has been filled since some tens of years by the so called Electrochemical Double Layer Capacitors (EDLC). A Ragone diagram of some currently available power sources is schematically and qualitatively displayed in Figure 2.1.

---

[1] Electric energy storage and conversion devices.

For completeness, also the fuel cells are mentioned in the Ragone diagram. They

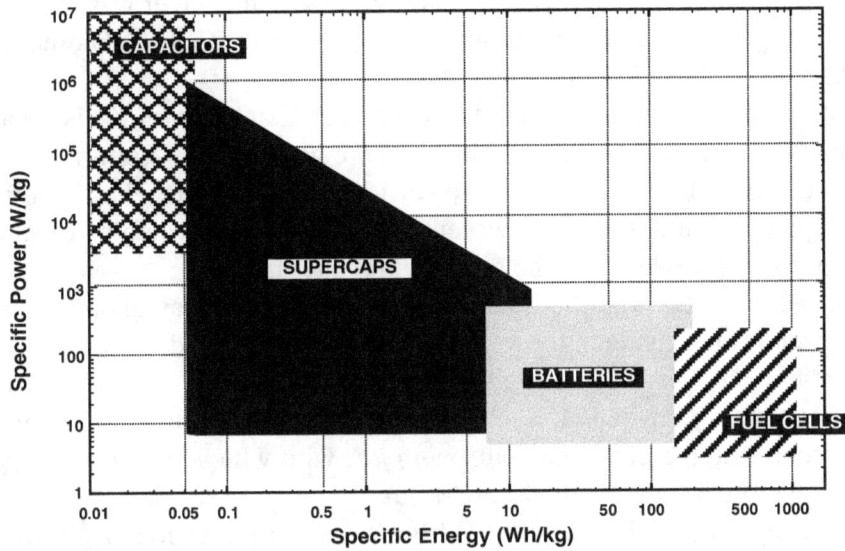

Figure 2.1: Ragone diagram comparing schematically and qualitatively Capacitors, EDLC (Supercaps, Supercapacitors® etc.), Batteries and Fuel Cells concerning power and energy density.

have the highest energy density and the smallest power density among all power sources considered here.

If no EDLC were available, and a capacitor with a particular energy was required for some specific application, a particular number of conventional capacitors would have to be applied in parallel to achieve the required energy.

As the EDLC have a higher energy density, they could replace the conventional capacitor arrangement with the result that less space or weight was used up.

There exists a number of promising electrode materials for EDLC, such as high surface area carbons (fibers, foams, aerogels, composites, glassy carbon), doped conducting polymers (polyaniline) and mixed metal oxides (ruthenium, iridium and tantalum oxides and, recently, perovskites) [4]. The most important material properties of EDLC electrodes are the specific capacitance (F/g) and the electronic and ionic resistivity (Ωcm), which eventually determine the specific energy and the specific power of the capacitor.

2.2 Fundamentals

2.2.1 The Electrochemical Double Layer

The phenomenon of electric charge storage in a double layer was observed first by Ewald Jürgen von Kleist, born on March 7, 1715 in Zeblin (Pommern) and died on August 24, 1759 in Frankfurt/Oder.
Kleist, who belonged to the family of the poet Heinrich von Kleist, was the dean of the cathedral and later the president of the yard court in Pommern.
Kleist electrified a nail inserted into (inside probably somewhat damp) a medical glass apparatus and received a violent electrical impact, when he touched the nail with one hand and keeping the glass in his other hand.
The effect was still amplified, when the bottle contained alcohol or mercury.
His observations, made in the year of 1745, were reported soon toward the physical society and found considerable attention.
One year later in Leyden/Holland the same attempt was made by Pieter van Musschenbroek, and therefore the *amplifying jar*, with which now many experimentalists began to work, was called *Leyden jar*.
Historically correct the name would be *Pommern jar* or *Kleist jar*. Soon the jar was plated with metal coatings and its strikes and flashes found much attention [5].

Whenever an electrode is inserted into an electrolyte, an electrochemical double layer [2] is built up at the interface between electrode and electrolyte. For metal electrodes, the surface has a negative charge which extends to the outer space and may attract positive charges, when present in the electrolyte. The physical origin for this phenomenon is the difference in the chemical potential

$$\mu_i = \mu_i^0 + R\,T\,ln\,a_i \tag{2.1}$$

for each component i with activity a_i present in a system of different phases I and II, which are in a direct contact with eachother:

$$\mu_i(I) \neq \mu_i(II) \tag{2.2}$$

An example of a simple system is illustrated in Figure 2.1, a copper rod being the electrode (phase I) and a copper sulfate solution being the electrolyte (phase II) [6, 7, 8, 9]. In general, the thermodynamical equilibrium condition

$$\mu_{Cu++}(solution) = \mu_{Cu++}(metal) \tag{2.3}$$

---

[2] Furtheron, the electrochemical double layer will be abbreviated in this thesis with EDL.

will not be fulfilled immediately after insertion of the electrode in the electrolyte. However, the equilibrium will be gained by a chemical reaction [10] on the electrode surface, either by an anodic dissolution of the metal or by cathodic coating from the solution,

$$Me^{z+} + ze^- \rightleftharpoons Me \qquad (2.4)$$

accompanied by an electrical potential difference φ. The condition for the equilibrium yields

$$\mu_i(I) + z_i F \varphi(I) = \mu_i(II) + z_i F \varphi(II). \qquad (2.5)$$

In the case of dissolution, the metal ions will charge the boundary layer (double layer) on the electrolyte side positively, and the electrode remains negatively charged. At the isoelectric point, the dissolution will be stopped by electrostatic forces. The equilibrium can be perturbed by applying an external DC potential with the result, that the charge of the double layer can be increased or decreased, or the sign of the charge can change. If the external applied potential does not exceed the electrolysis potential of the electrolyte and the dissolution potential of the electrode material, no faradaic current will occur. In this case, only the EDL will be recharged. This case will be focussed on in this thesis. The similar phenomenon occurs at solid-solid interfaces of semiconductors, for example. It is an important fact that the EDL has no electric conductance, so that a potential drop, which accounts for an electric field, occurs over the EDL when a voltage is applied.

2.2.2 Working Principle of Supercapacitors

Electrochemical double layer capacitors (EDLC) take advantage of an energy storage mechanism in the electrochemical double layer.
Consider an electrode inserted in an electrolyte and the double layer at the interface between electrode and electrolyte, as displayed in Figure 2.3.
 The potential relations in the schematic are by far oversimplified, but serve well for the understanding of the working principle for EDLC.
The electronically fairly well conducting electrode has a constant potential at a particular value.
The electrolyte has a potential different from the electrode. At the interface (the double layer), a potential gradient exists which yields an electric field.
The double layer has a thickness of around a monomolecular or monoatomar layer of around 1 to 2 Å.
The electric field is

$$\vec{E} = -\epsilon_r \epsilon_0 grad\phi = -\epsilon_r \epsilon_0 \frac{d\phi}{dx}\hat{e}_x \qquad (2.6)$$

2.2. FUNDAMENTALS

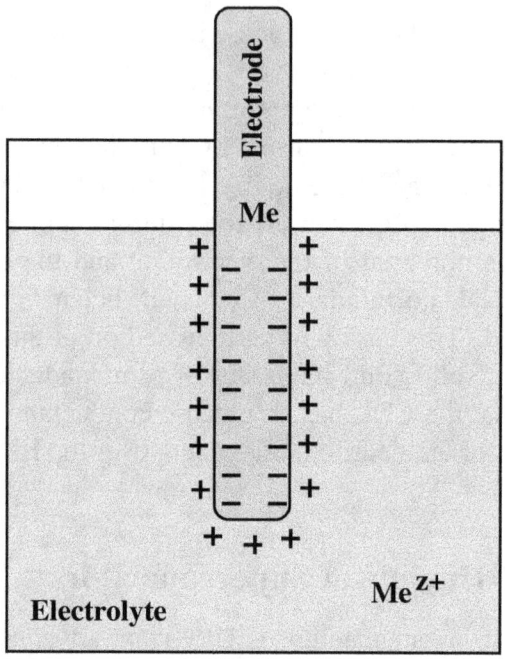

Figure 2.2: Potential difference between electrode and electrolyte and electrolytic double layer at the phase boundary metal/electrolyte: $\mu_{Me^{z+}}$(metal) < $\mu_{Me^{z+}}$(solution).

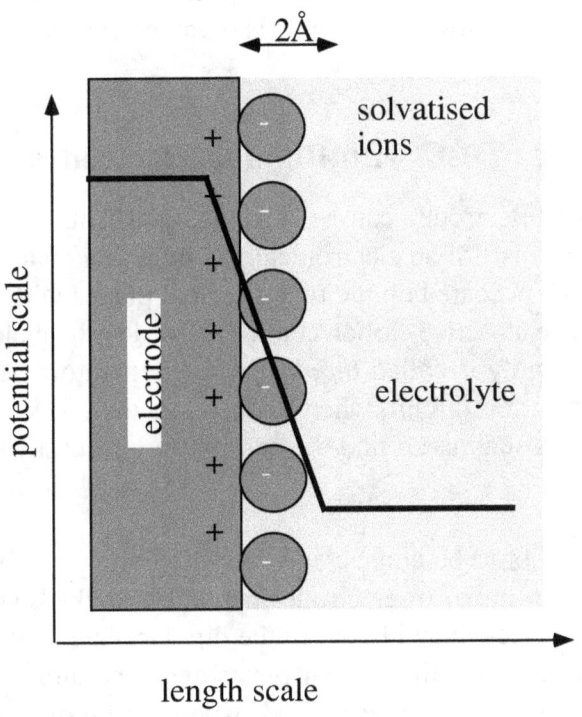

Figure 2.3: Sketch of potential at electrode, double layer and electrolyte.

2.2. FUNDAMENTALS

and the energy density in the field is

$$W = \int \vec{E} d\vec{x} = \frac{1}{2}\epsilon_r \epsilon_0 \vec{E}^2, \qquad (2.7)$$

ϵ_r being the relative dielectric constant (water: ≈ 10) and ϵ_0 being the electromagnetic field constant ($8.854 \cdot 10^{-12}$ As/Vm).
The potential difference in an electrochemical cell may not exceed a voltage of around 3 Volt, if the electrolyte is an organic electrolyte (such as acetonitrile). At higher voltages the electrolyte will be decomposed. If an aqueous electrolyte is used, the decomposition voltage is around 1 Volt. With above data, an upper limit for the energy density of around 1 kJ/cm^3 is obtained for the double layer.

2.2.3 Design of EDLC with Bipolar Electrodes

EDLC with bipolar electrodes can be designed as displayed in Figure 2.4. A bipolar electrode consists of an electronically conducting plate or foil (made from metal or GC), which is coated on the two principal planes with an ion conducting material, for instance an active carbon coating soaked with an electrolyte (ion conduction). This assembly is called *bipolar plate*. Separators between the bipolar plates avoid electric short circuit, when stacked together. Two end plates serve as current collectors. A sealing around the end plates is necessary to avoid leakage of the electrolyte.

Bipolar plates need not to be connected with wire contacts, which is particularly useful when a large number of electrodes has to be stacked, i.e. in high voltage applications. Resistances in and between the bipolar plates will become very low (ionic and electronic) when the bipolar plates and separators are very thin. The current distribution [11] is also very good, when the bipolar plates are thin and their distances are small.

We focus on EDLC based on Glassy Carbon (GC) electrodes with a high power density. When monolithic GC plates are oxidized, a sandwichlike structure with a porous film on top and bottom of the non-oxidized GC is obtained. While the porous film on each side can be filled with electrolyte and therefore is ion conducting, the non-oxidized part in between is only electronically conducting and acts as a current collector. Such samples can be utilized as monolithic bipolar electrodes for EDLC [12].

GC was selected as an electrode material for EDLC for several reasons:
(i) Activated GC has a high internal surface area accessible for liquid electrolytes

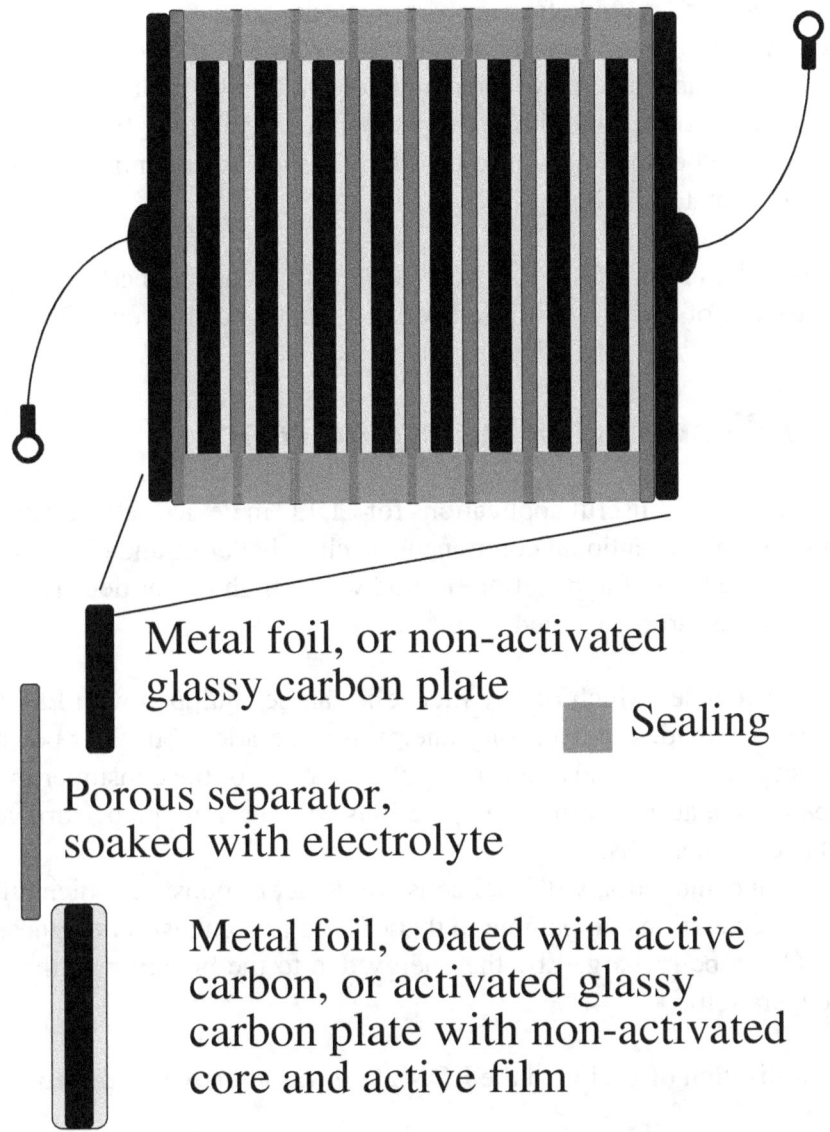

Figure 2.4: Sketch of a capacitor stack with bipolar electrodes.

2.3. APPLICATIONS FOR SUPERCAPACITORS

and therefore a correspondingly high double layer capacitance [4];
(ii) GC has a good electronic conductivity (200 S/cm);
(iii) The monolithic approach allows for a very low internal resistance [13], because grain to grain contacts and grain to current collector contacts are omitted - in contrast to electrodes made from powder material [14];
(iv) Although carbon is not thermodynamically stable at electrochemical potentials [15], GC is kinetically very stable and does not undergo any chemical reactions in many electrolytes and thus exhibits long term stability [16].
(v) Compared to other materials used for EDLCs, such as ruthenium oxide, GC is a fairly low cost material [4].

With the monolithic approach, a 5 Volt stack with a capacitance of 0.45 F and a serial resistance of 26 mΩ was developed [12], as was displayed in Figure 1.1.

2.3 Applications for Supercapacitors

There exist a variety of useful applications for EDLC in devices already available, but equipped with conventional components such as batteries and capacitors. Depending on the kind of application, EDLC with a high power density or a high energy density have to be selected.

- Electric vehicles which run by fuel cells can be equipped with EDLC. The fuel cells provide the necessary energy for operation, but their power density is too low for typical driving cycles of a car for the consumer market. In particular, accelerations are operations which cannot be performed comfortable by fuel cells.
  EDLC in conjunction with fuel cells would act as boosters which allow for a fast acceleration comparable to that of regular combustion engines.
  EDLC can be recharged by the energy due to the breaking of the vehicle (recuperation).

- The utilization of EDLC in the LOK 2000 was already mentioned.

- EDLC can be used for power back up systems such as in computers. In case of any event which affects the power supply of the computer (or any other device with volatile memories), the EDLC would provide electric energy sufficient for the computer to run down safely. In particular, the EDLC meets the requirement that the necessary voltage is maintained immediately (high power density) after the event. As the back up is carried out in only a few seconds, the energy required is not so high.

- Nonstationary communication systems are operated with rechargeable batteries sometimes (lithium ion batteries etc.). During transmission they operate with pulsed power, which considerably reduce the cycle life of the batteries. When an EDLC is used as a booster for pulsed operation, the cycle life can be enhanced.
  This is also valid for batteries in electric vehicles.

- EDLC can be utilized when the power quality needs to be improved. The electricity in some countries (USA, for instance) is very spiky and therefore can destroy devices.

2.4 Capacitor World Market

The information presented in Figure 2.3 is based on a publication of A. Nishino (Matsushita Electric Industrial Co. Ltd., Osaka, Japan) in Journal of Power Sources 1996 [17] and on statistical data [18, 19]. The worldwide market for capacitors (including traditional capacitors and EDLC) in 1993 was around 12.3 billion US$. In 1994 the EDLC contributed to 9-10 billion Yen to the whole capacitor market,

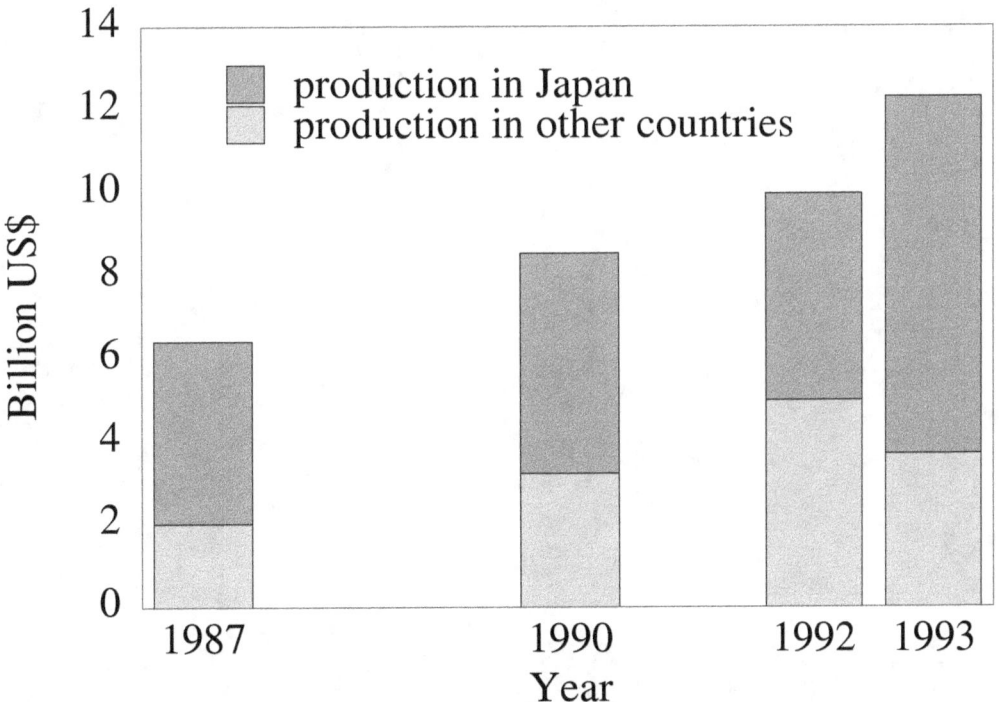

Figure 2.5: Capacitor market share in the last decade.

2.4. CAPACITOR WORLD MARKET

and it is believed that it will grow to 100 billion Yen by the year 2000.

Now the EDLC research and development are in a consolidation phase, probably because previous high expectations gave way to more realistic assessment of the potential of the EDLC.

Especially research groups whose original business was in battery research and who attempted to develop EDLC with their know how, draw back in order to improve batteries towards higher power density.

Therefore there is less need now to improve the supercapacitors towards a higher energy density, because they have to compete with novel batteries.

At present the EDLC have the highest potential in the development of capacitors with a high power density.

Chapter 3

Glassy Carbon

Figure 3.1: Interlaced hexagon. Symmetry Drawing 133. By M.C. Escher (1967).

Carbon exists in a variety of modifications, the most prominent of which are diamond and graphite.
They are the purest species of carbon in terms of crystal structure.
Diamond consists of carbon atoms tetrahedral bounded to accommodate all electrons without distortion, the carbon atoms therefore fitting into the classical diamond type cubic lattice, which can be regarded as two interpenetrating cubic

lattices based on points with Miller indices (0 0 0) and ($\frac{1}{4}$ $\frac{1}{4}$ $\frac{1}{4}$) [16].

Graphite consists of sheets of trigonally coordinated carbon atoms in the sp_2 state, and the sheets are stacked in a hexagonal $ABA\ldots$ sequence.

The covalent bonds in the basal plane are extremely stiff and strong, while the bonds between the basal planes arise from the rather weak van der Waals interactions so that the crystal can be sheared easily [16].

In recent years other species of carbon were discovered: first the spheric so-called Buckminster Fullerenes [20], later the carbon nanotubes [20], the carbon atoms all of which are bounded by covalent interactions.

Besides these modifications, which are highly crystalline and anisotropic, also less *pure* and amorphous, but nonetheless important kinds of carbons exist such as coals, char and cokes.

So carbon atoms arrange in a whole variety of structures. Figure 3.1 displays a drawing of the famous belgian artist M.C. Escher [1]. The hexagon is a pattern which is obeyed by the graphite. The *interlaced hexagon*, an idea by Escher, is a pattern which is maybe physically impossible or improbable to be obeyed by atoms. But it should serve as an animation for the reader of this thesis.

Glassy Carbon (GC) [21, 22] is another modification of carbon which appears shiny as glass with a black colour. GC has a pronounced mechanical strength which is similar to glasses. Some physical data of GC as purchased from HTW Hochtemperatur Werkstoffe GmbH, Thierhaupten/Germany, are listed in Table 3.1.

Investigated with X-ray diffraction, diffraction peaks of GC occur at the positions where graphite peaks are expected. However, peaks are broad and therefore crystallites are very small. Therefore GC cannot be regarded as a really crystalline material.

Unlike the interlaced hexagon, the glassy carbon (micro-) structure is regarded as a randomly oriented arrangement of graphite crystallites as schematically displayed in Figure 3.2. GC consists of stacks of graphene sheets with lateral extensions (usually denoted L_a) between around 25 Å and several 100 Å, and stack heights (usually denoted L_c) between 2 or 3 up to several dozen sheets, depending mainly on the heat treatment temperature (HTT) [16, 23]. Figure 3.3 displays schematically the microstructure of GC which was pyrolyzed at a higher HTT. The crystallites of high HTT GC are larger than those of low HTT GC, as depicted in Figure 3.2.

The graphene stacks enclose voids, so-called micropores, with only a few nanometer in size. While the graphene sheets are bound together by van der Waals interactions, the crystallites are probably bound by covalent forces, which account for

---

[1]The authors attention was focused on M.C. Escher during his undergraduate studies in crystallography with Prof. Th. Hahn, RWTH Aachen.

Figure 3.3: Schematic sketch of the microstructure of GC with a low HTT. After high HTT, the crystallites are large extended (picture after Jenkins and Kawamura [16]).

the very robust mechanical properties of GC.

Due to the presence of the voids, the raw density of GC is rather low (around 1.5 g/cm^3), compared to graphite (2.26 g/cm^3) [24, 25].

As the voids are closed and separated to each other, GC is impermeable to gases and liquids [26, 22]. However, the pores can be opened by an appropriate oxidation process (*activation*; oxidation and activation refer in this work to the same process.), such as gas oxidation [27] or electrochemical oxidation [28, 13]. The pores then become accessible to gases and liquids.

The resistance and capacitance of activated GC depend on activation parameters such as the activation temperature, activation time and concentration of the oxidant. Nitrogen Gas Adsorption, Electrochemical Impedance Spectroscopy, Scanning Electron Microscopy, Small Angle X-ray Scattering and X-ray Diffraction were applied to characterize the activated GC.

GC can be prepared by pyrolysis of polymers (phenolic resins or furfuryl alcohol) under inert atmosphere at heat treatment temperatures (HTT) between 600°C and 3000°C [16].

Above temperatures of 300°C, the polymers loose their non-carbon contents (hydrogen, for example) as gases and undergo a process, which is termed carbonization.

The material changes from the starting polymer to a form of carbon. If the temperature of pyrolysis exceeds 1000°C, almost any non-carbon content is eliminated, and the form of carbon obtained is called polymeric carbon.

Depending on the structure of the precursor material and the further treatment of the carbon, either activated charcoal with enormous available internal surface areas, or glassy carbon with effective barriers impermeable even to helium, can be obtained.

3.1 On Activated Glassy Carbon

The difference between glassy carbon in its natural state and activated carbon is that the pores are separated and closed against eachother, whereas the latter has intersected and connected pores.

In fact, in former case one should avoid the name pores and use instead *voids* for the empty space between the graphene sheets.

If these voids become opened and connected by the oxidation process, they become accessible to gases and liquids and thus may be called pores.

There is a difficulty in describing the structure and morphology of porous media, especially if the matrix is disordered and the pore network is randomly built up.

3.1. ON ACTIVATED GLASSY CARBON

Similar to glasses or amorphous materials, in contrast to crystals, which are highly ordered materials, some kind of *correlation length* must be introduced to characterize the material.
The correlation length of disordered materials is the pendant to the lattice parameter of crystals.
Another measure for the characterization of porous media is the pore size distribution, which can be obtained from adsorption measurements (using gases or liquids as adsorbents) and also from small angle scattering experiments (using neutrons, x-rays, or light).
However, the pore size distribution alone still may be insufficient to characterize the material, if *connectivity* of the pores is not taken into account.
The connectivity of pores is an important feature of electrodes inserted in electrolytes, because in our case the interface between solid electrode and liquid electrolyte has to be as large as possible.
In the case of supercapacitor electrodes, this feature amounts in a valuable amount of double layer capacitance per sample volume.
The geometrical topology of the pore space describes the extent of its connectivity.
Essential characteristic on the pore space is

- N_V, the number of isolated areas inside the pore space which have no connection to eachother (0. Betti number).

- C_V, the number of redundant connections inside the pore space, i.e. the maximum number of connections which may be cut off, although no additional isolated area is created (Genus, or 1. Betti number).

A global topological measure of the pore space is for instance the so called *Euler-Poincaré-Characteristic* ϵ:

$$\epsilon = N_V - C_V. \qquad (3.1)$$

For large positive values of ϵ, the networking may be regarded as poor. For large negative values of ϵ, the networking may be regarded as extensive.
In the case of unactivated GC, the 0. Betti number N_V describes the number of isolated pores: voids.
As nearly all the pores in unactivated GC are closed, this number is very high.
The Genus C_V is very small or almost zero, because no pores are connected. Moreover, no redundant connections exist.
Thus ϵ of non-activated GC is a large positive number, i.e. the number of pores, or simply, the porosity.

In contrast, this procedure may also be applied on the GC crystallites, instead on the pores. So the GC crystallites can be regarded as being embedded in an empty-space matrix [2].

Then the 0. Betti number N_V^* concerning the GC crystallites is in the same range as the number of pores. Roughly one expects $N_V^* \approx N_V$.

However, the crystallites are interconnected and may have redundancies. Therefore the Genus C_V^* is a positive number, and moreover: $C_V^* > C_V$.

Finally we have $\epsilon^* > \epsilon$.

In this thesis it will be shown that the number of pores decreases during activation by coalescence of smaller pores to larger pores [3]. A decreasing pore number means that the 0. Betti number also decreases. The decrease of N_V becomes even more pronounced because the remaining pores become interconnected.

Furtheron, the pores become multiply interconnected and therefore the Genus C_V increases.

So the characteristic ϵ is getting smaller during activation. Anyway, qualitatively following relation is valid: $\epsilon_{act.} < \epsilon_{non-act}$.

Experimentally, ϵ can either be determined by determining N_V and C_V directly through a counting procedure [31], which requires a certain number of serial sections of the material under investigation.

Only two serial sections are required for the determination of ϵ for the so-called disector method [32, 33, 34].

Recently soil sciences took advantage of a new method of obtaining three dimensional representations of soil samples and their pore topology by techniques such as serial sectioning and tomography [35, 36, 37] and subsequent digital imaging processing.

This technique is not yet so well developed that structures significant for electrode properties can be resolved, i.e. pores in the range of nanometers. The microfocus beamlines at synchrotron radiation sources of the state-of-the-art generation can resolve structures spatially to a limit down to 100 nm [38].

Therefore indirect methods (diffraction and scattering) are necessary to characterize the porous structure of GC. In the next chapter a method for the determination of the film thickness will be introduced first.

The closed pores of GC can be opened by oxidation processes. Probably part of the interlinking atoms, which glue together the small graphite crystallites by covalent forces, are removed by the oxidation.

---

[2] in analogy to Babinet's principle in optics [29]

[3] The author refers to the falling cards model, as proposed by W. Xing et al. [30]

3.1. ON ACTIVATED GLASSY CARBON

As a result channels are left behind, which serve as diffusion channels for further oxidation in deeper regions of the GC.
Several ways of GC activation are described in literature, and some of them will be introduced briefly in this thesis.
Gas phase (thermal) oxidation with, for instance, simply with air as an oxidant in a muffle furnace, was carried out in this thesis work.
Electrochemical oxidation in 3 molar sulfuric acid was also carried out mainly prior to this thesis by coworkers of the authors, but also by the author himself. Also perchloric acid was used as an oxidant.
Another way to activate GC is to use sulfuric acid at high pressure and temperature in an autoclave, as patented by Siemens.

In the supercapacitor group at PSI it was found that the electrochemical oxidation is an appropriate activation process for GC pyrolyzed at high temperatures (around 2000°C, while the thermal gas phase oxidation is appropriate for the activation of GC pyrolyzed at temperatures around 1000°C.

3.1.1 Thermal Activation of Glassy Carbon

Whenever in this thesis the terminus *activation* is used, we mean a process which acts to increase the internal surface area of carbon for the enhancement of its sorptive properties.
Very often oxidation is such a process, which can be carried out in various ways, such as wet chemical oxidation (in boiling nitric acid for instance), electrochemical oxidation (as an electrode inserted in an electrolyte, which is some kind of anodic dissolution [28, 13]), and thermochemical oxidation [39].
Thermochemical oxidation means that the carbon can react with an oxidant (usually a gas) at elevated temperatures, so that the internal surface area accessible for liquids and gases increases drastically.
This process will be called furtheron *thermal activation*. The thermal activation of GC was first reported in a patent of Siemens [39] in 1980 [4].

The activation of carbons, such as coals and graphites, is in general well understood because of their importance in chemical industry and technology [40]. However, not much work has been reported on the activation of GC.
In recent years GC and other hard carbons received attention as electrode materials in electrochemical energy storage research [41, 42, 30].

---

[4]Although the Siemens research group succeeded in developing an EDLC based on GC, no commercial product turned out of their project, because the product would have been too expensive (" ... die Kaufleute haben das totgerechnet", quotation by Dr. M. Waidhas, Siemens.).

The compact, electronically well conducting GC matrix of activated solid samples shows a very good frequency response of the capacitance [43], compared to electrodes made from powder materials [14].
However, the resistance and capacitance of activated GC depend on activation parameters such as the activation temperature, activation time and concentration of the oxidant.
Finally, when compared to other materials for EDCs such as ruthenium oxide, GC is a fairly low-cost material [4].
GC monolithic samples prepared like this are very suitable as electrodes for so called supercapacitors, a novel class of energy storage devices with a very high energy and power density [12, 3].

Thermal activation of GC can be performed easily by using a furnace as a reactor and air as a reacting oxidant.
At temperatures between 400°C and 600°C, within less than one hour GC with a specific surface area of around $1000 m^2/g$ can be obtained.
The proper choice of activation time and temperature to achieve maximum values for internal surface area and capacitance depends on the type of GC used and on the concentration of the oxidant.
Compared with the other ways of activation mentioned above, thermal activation is a rather clean and cheap process and therefore favourable.

The thermal activation of GC for supercapacitor applications experienced revival in the Swiss Priority Program on Materials Research (PPM) in 1995, when a collaborative project was raised from Asea Brown Boveri (ABB), Leclanché S.A. and Paul Scherrer Institut.
This project was judged repeatedly successful and promoted finally into a KTI project (Kommission für Technologie und Innovation) under leadership of ABB.

Chapter 4

Experimental

4.1 Sample Preparation

Samples as received by the manufacturer [1] were rinsed in deionized water and dried with a paper. With a diamond saw rectangular pieces of usually 1 cm×2 cm in size were obtained. Depending on the measurement technique which had to be applied later, smaller or larger samples were necessary. Then these samples were kept in boiling nitric acid (HNO_3) for 15 minutes, and then again rinsed in deionized water and dried in a paper tissue.

For thermal activation, an electric muffle furnace (Nabertherm®, model L9/SH) with a volume of 15 liter was used. The temperature of the furnace could be electronically controlled and stabilized within a range of accuracy of ± 5 K. For use with various reaction gases, the furnace could be supplied with external gases by a junction. If not stated otherwise, this junction was interrupted and open, so that just the air from outside could circulate into the furnace. The junction had a diameter of around half a centimeter.

The GC concerning this thesis was thermally oxidized in a furnace reactor with air as an oxidant. Some samples were also oxidized with other gases or gas mixtures (O_2 [2] or CO_2 [3] at different concentrations). If other oxidants than air were used, it will be noted accordingly.

Samples were mounted on a sample holder block made from high porous silicon oxide. Slits to hold the samples were cut in the block with a knife. Samples on the sample holder were brought into the pre-heated and temperature stabilized furnace. Care was taken that the temperature decay did never exceed 5 K. After the required time for activation the sample was removed from the furnace to cool

---

[1] HTW Hochtemperatur Werkstoffe GMBH, Gemeindewald 41, Thierhaupten, Germany
[2] Olivier Merlot, Université de Fribourg, Internship at PSI in summer 1997.
[3] Dr. Martin Bärtsch, PSI.

down at ambient atmosphere.

Sometimes it was necessary to prepare series of samples activated at one specific temperature. Therefore several samples were brought into the furnace at the same time, and after a specified time the first sample was removed to cool down, while the other samples remained in the furnace for further activation. Care was taken that supplying the furnace with the sample happened quick enough (less than 5 seconds) so that not much cool air from outside could penetrate the furnace and not much hot air from inside the furnace could escape. This procedure was only applied to samples with an activation time longer than 30 minutes, so that a possibly cooling down of the samples by the numerous opening of the furnace door could be neglected. When a series of samples with shorter activation times was required, always single samples were activated.

Glassy Carbon powder was thermally activated as follows. The GC powder was suspended in concentrated ethanol, then poured in a dish with 25 cm diameter and subsequently waved with circular motions so that the suspension was spread throughout the dish and covered the walls of the wish with a fine film. Soon the ethanol evaporated and a thin film of GC powder remained on the dish wall. The powder was removed from the dish with a fine brush and collected for measurement purposes.

4.2 Thickness Determination

4.2.1 Determination of the Sample Thickness

Sample thicknesses were determined with a micro screw gauge, with a micro profile detector (*Stylus MT 12B, Heidenhain*) and with a *TOPCON 60* Scanning Electron Microscope (SEM).

The micro screw gauge had two plates with 2 cm diameter, between which the sample was placed for measurement. This allowed a quick determination of sample thicknesses with a limit of accuracy of $\pm$ 5 μm. Samples with an inhomogeneous thickness could not be measured accurately, because only the maximum thickness on a length scale smaller than the plate diameter could be obtained.

The Stylus MT detector allowed a thickness determination with an accuracy not less than 1 μm. Difficulties arose during the measurement of bended samples, as in the case of thin GC sheets, because pressure had to be applied on the detector tip to force samples to become flat and plane, when necessary.

The most accurate thickness determination was possible using the SEM, which had the particular advantages that (i) printed images of the sample were available for further evaluation and (ii) even bended samples could be measured precisely.

4.2.2 Determination of the Film Thickness

The active film thickness was determined by Scanning Electron Microscopy (SEM). Before applying SEM, the sample were soaked with a contrast medium (*silver nitrate*, $AgNO_3$), which facilitated the thickness determination essentially.
The procedure of thickness determination was as follows:
After removal from the reaction furnace, the monolithic samples were immersed in a glass with saturated $AgNO_3$-solution at a temperature around 80 °C for 15 minutes. Then the samples were removed from the solution and laid on a flat glass plate and evacuated in a drying furnace at 80°C for around one hour. Then the samples were removed from the drying furnace and broken either wrapped in a paper tissue or across the edge of a glass plate. The broken samples then were mounted on a sample holder for electron microscopy, and the fracture area was investigated with electrons at an accelerating voltage between 10 and 20 kV.
Activated samples in the SEM exhibited a bright envelope and a dark center part. It was assumed that the bright area reflects the active film, impregnated with dried $AgNO_3$, and the center part the non-activated bulk GC material. The brightness of samples investigated with SEM is a measure for the electronic conductivity of the samples. Insulating samples appear bright, because they cannot be grounded. Conductors like metals appear dark. Variations in the brightness on the samples can be interpreted as variations in the conductivity. Some of the samples were investigated with Auger Electron Spectroscopy (AES) in order to verify that the bright region in fact is impregnated with the silver salt. Applying a line scan across the sample with sensitivity for energy of the electrons arising from silver, it could be shown that indeed the bright area of the sample is impregnated with silver, but the dark part not.
The mapping of the samples by the SEM was not true for length scales. So a calibration of the SEM with a patterned reference sample (microstructured silicon wafer plate with known periodicity) was carried out. From printed SEM images the film thickness could be measured with a ruler.
This method was applied to a series of samples activated at a specific temperature for different times. Only thin (60 μm and 100 μm) GC samples of the Sigradur®K GC could be measured easily.
The controlled breaking of GC samples remained a problem at last. With thick samples of 1 mm thickness always a shell fracture pattern occurs which makes film thickness determination difficult, because reflexion effects from the SEM cannot be distinguished from morphological effects of the sample. It was tried to break the samples in liquid nitrogen (LN_2). But results were poor as well.

4.3 X-ray Diffraction

X-Ray Diffraction (XRD) was performed with a Philips X-Pert Diffractometer, having a copper target X-ray tube (CuKα, λ=1.54056 Å). Diffractograms were automatically recorded using the Philips data acquisition software. The limit of angular resolution was 0.05°. Evaluation of the diffractograms however was performed manually using a standard laboratory data analysis program (KaleidaGraph, ©SYNERGY SOFTWARE). All experiments were carried out in the reflexion mode. The solid samples were placed on the sample table on a glass support. The section of the incident beam on the sample was 5 mm × 5 mm. Care was taken that the altitude of the sample surface was adjusted so that the diffraction peaks obtained from a silicon standard sample were in accordance with the silicon reference pattern JCPDS 27-1402 [44]. To calibrate the 2Θ scale and to verify for the positions of diffraction peaks, from time to time the surface of the GC samples to be measured was covered with a silicon powder standard material. In the case of GC, the (002) peak is of considerable interest, because this is usually the only well developed peak of GC and other carbons. See Figures 4.1 and 4.2 for the XRD reference patterns of rhombohedral and hexagonal graphite [44].

The measurement of solid monolithic GC samples as received from the manufacturer was particularly difficult, because the thickness of the samples affected the peak height and minor the peak positions [4]. When samples had a different thickness, the height of the sample surface also changed. As a result, an error in the peak position occurred. The other reason was actually the amount of GC material transmitted and reflected by the X-ray beam.

To quantify the influence of this effect, XRD diffractograms were recorded for single GC sheets and for stacked sheets. The sheets were stacked from below, so that the sample surface height with respect to the diffractometer geometry never changed. Then the (002) peak height and position as a function of the amount of GC were measured.

This investigation is of particular importance when activated GC samples are measured, which have experienced a substantial burn-off during activation and therefore are thinner than the non activated samples. The altered sample thickness therefore must be taken into account in XRD measurements. In order to quantify the thickness effect, the *sample thickness* was varied by stacking thin K 800 samples with each 60μm thickness (First one, then three, five and seven sheets) from below the sample table, so that the sample surface height remained constant.

---

[4] It was found that samples with different thickness have different structure. See Section 6.5.2 for details

4.3. X-RAY DIFFRACTION

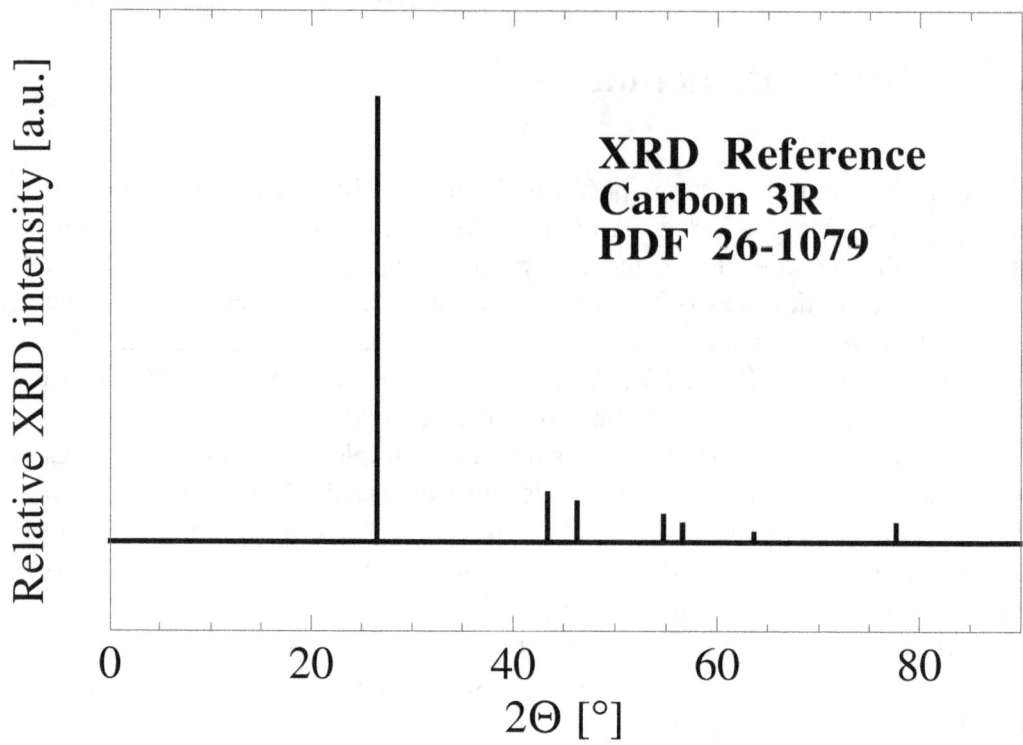

Figure 4.1: Reference diffraction pattern of Graphite, Powder Diffraction File 26-1079 from ICDD [44].

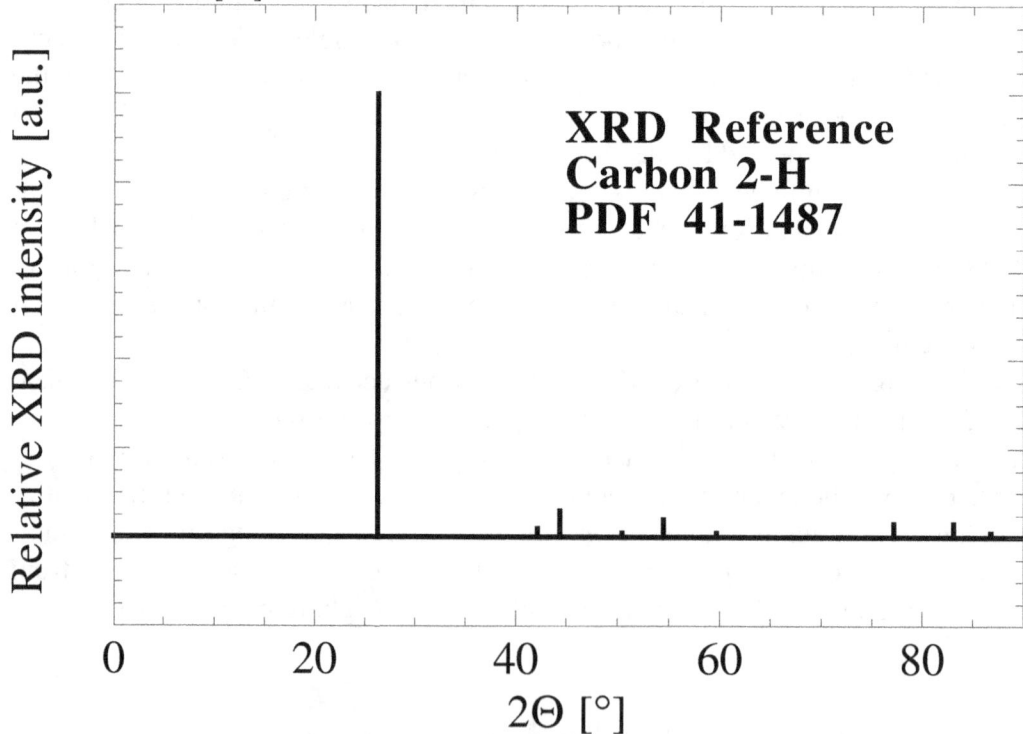

Figure 4.2: Reference diffraction pattern of Graphite, Powder Diffraction File 26-1079 from ICDD [44].

4.4 Small Angle X-ray and Neutron Scattering

4.4.1 Small Angle X-ray Scattering

Small Angle X-ray Scattering (SAXS) experiments were carried out at the Hamburger Synchrotronstrahlungslabor HASYLAB (Deutsches Elektronensynchrotron DESY, storage ring Doris III) at the JUSIFA Beamline, which is operated by the Institut für Festkörperforschung, Teilinstitut Streumethoden, Forschungszentrum Jülich GmbH, Jülich, Germany [45, 46].
The synchrotron radiation was monochromized using two silicon crystal monochromators. The energy scale was calibrated by x-ray absorption spectra obtained from metal samples, such as iron and platinum. A two-dimensional position sensitive wire detector (PSD) with 256 × 256 pixels, sized 18 cm × 18 cm was used to record the scattered intensity.
The scattering curves were obtained with photons of solely 11 keV or with 7.1 keV and 13 keV energy. To cover a sufficiently large range of Q-vectors for a measurement with proper resolution, the detector could be positioned in each of 4 different distances from the samples, the shortest distance being 935 mm and the longest being 3610 mm. The scattering curves were obtained either at 11 keV for both geometries or by 7.1 keV for the long and 13 keV for the short geometry.
The samples were measured in a vacuum chamber (p $\approx 10 \times 10^{-5}$ mbar), mounted on a sample holder which could carry 5 samples at the same time. For calibration a GC reference sample, whose scattering cross section is known, was measured. Therefore, scattering curves in absolute units [*electron units*, e.u.] could be obtained.

The data reduction of the two-dimensional scattering data was performed using specific evaluation software on site, which is explained in Appendix C (Program *Auswert*) to this thesis. The output of *Auswert* is a one-dimensional data set with one column for the scattering vector (Å^{-1}), a second column for the scattering cross section (e.u./cm^3, the measured intensity), and a third column called error matrix, which denotes the absolute error in the second column.

Determination of the Scattering Contrast

For further quantitative analysis of SAXS scattering curves, information on the scattering contrast is necessary. It is the linking part between the intensity of measured curves and the quantities which are to be obtained by the analysis procedure. The scattering contrast of a particle in a specific matrix material is defined as

$$\Delta n_f = n_{matrix} \cdot f_{matrix} - n_{particle} \cdot f_{particle}, \qquad (4.1)$$

4.4. SMALL ANGLE X-RAY AND NEUTRON SCATTERING

where n denotes the number density of the matrix material [atoms/nm^3] and f is the atomic form factor [f $\cdot \sqrt{e.u.}$][5]. The molar mass of carbon is 12.011 g/mol. The number density of the GC matrix material is given by

$$n_c \left[\frac{atoms}{cm^3}\right] = \rho_x \left[\frac{g}{cm^3}\right] \times \frac{6.02 \cdot 10^{23} \left[\frac{atoms}{mol}\right]}{12.011 \left[\frac{g}{mol}\right]}. \qquad (4.2)$$

The X-ray density ρ_x of the GC sample can be calculated from X-ray data according to [48]

$$\rho_x = (3.33538/d_{002}) \times 2.268 \, g/cm^3. \qquad (4.3)$$

From equation 4.2 follows that the scattering contrast was obtained from the X-ray density, as measured in the XRD measurements. As GC contains voids rather than particles, equation 4.1 simplifies and the scattering contrast of GC therefore is given by

$$\Delta n_f \left[\frac{atoms}{cm^2}\right] = n_c \cdot f_c = n_c \cdot 6 \cdot r_e = \rho_x \cdot 8.474 \cdot 10^{10} \left[\frac{atoms \, cm}{g}\right]. \qquad (4.4)$$

Subtraction of bulk material signal

The partially activated samples, in contrast to non-activated and fully activated GC, consist of both active and bulk material. It is important, therefore, to correct the SAXS intensities for the bulk signal. The primary intensity I_0 is extincted by a sample with thickness D due to absorption (exponential decay), but at the same time is enhanced by scatterers, the number of which is linear in sample thickness. The scattered intensity I_{SAXS} is represented by the following relation:

$$I_{SAXS} = I_o \cdot A \cdot \frac{1}{V} \cdot \frac{d\Sigma}{d\Omega} \cdot D e^{-\mu D}. \qquad (4.5)$$

A is the sample area, V its volume, $d\Sigma/d\Omega$ is the differential scattering cross section, and μ is the linear absorption coefficient for carbon (2.514/μm at 11 keV energy). For $D = 1/\mu$ (approximately 300 μm), the scattered intensity has a maximum value. Below this sample thickness, the linear contribution is dominant. As the sample thickness is below 60 μm, we may apply the following linear subtraction of the contribution of the bulk material to the whole scattering signal. The total scattered intensity is the sum of one bulk contribution and two film contributions, each according to the relative thicknesses, as shown in the schematic

---
[5]e.u.: electron unit, Thompson scattering cross section of the free electron with a scattering length of r$_e$=2.818$\cdot$ 10^{-13}cm; f is the common value found in X-ray data tables [47]. For carbon at 11 keV: $f = 6$.

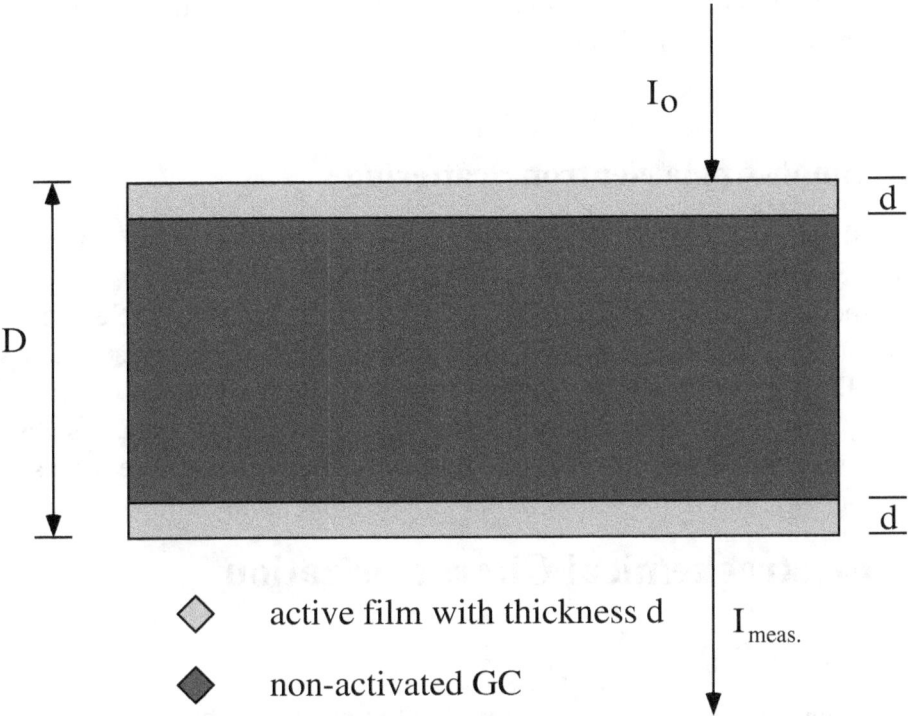

Figure 4.3: The sketch displays schematically an activated GC sample with overall thickness D enveloped by active film with thickness d on each side. The beam with primary intensity I_0 penetrates the active film, then the non-activated GC with thickness $D - 2d$ and finally the active film on back side of the sample. The beam therefore carries information from film and bulk material.

of Figure 4.3. The thickness of the film on each face of the sample is d, thus the signal intensity of the film is given by

$$I_{Film} = \frac{D}{2d} \cdot I_{SAXS} + (1 - \frac{D}{2d})I_{bulk}. \qquad (4.6)$$

Within this correction procedure, it was neglected that the interface between active film and non-activated bulk material could contribute to the scattered intensity. Such a contribution would have the largest influence on the data from samples with a small active film thickness.

4.4.2 Small-Angle Neutron Scattering

The GC sheets of SIGRADUR®G with 46 microns thickness were investigated with SANS, using neutrons with a wavelength of 8 Å.
SANS experiments were carried out at the Swiss Spallation Neutron Source (SINQ). Two-dimensional scattering patterns were obtained with a position sensitive wire detector. During measurement, samples were kept under a vacuum of 10^{-5} mbar. Scattering curves were obtained by radial averaging of the raw data. As absolute values in [barn/atom] were not yet available, only relative values are discussed

4.5 Electrochemical Characterization

4.5.1 Cell Setup

Electrochemical experiments were carried out in an electrochemical cell as commercially available from METROHM®. A schematic of the cell arrangement is displayed in Figure 4.4.
In all experiments, 3 molar sulfuric acid was used as an electrolyte. The 3 molar sulfuric acid was obtained from concentrated sulfuric acid (98%, pro analysi, Merck) and deionized water (ELGASTAT ® water purifier, water quality always better than 10 MΩcm). The electrolyte was not deaerated.
If it is otherwise stated, experiments were carried out at ambient temperatures of around 21°C.
In all experiments, a saturated Calomel electrode was used as a reference electrode. An untreated, clean Glassy Carbon (GC) disc of 1 mm thickness and 5 cm diameter was used as a counter electrode. The samples were contacted with a brass sample holder and covered with LACOMITE® varnish (LACOMITE VARNISH, Agar Scientific Ltd., Stansted, Essex, UK), with the exception of a free area used for the measurements, the latter being the only GC area subsequently

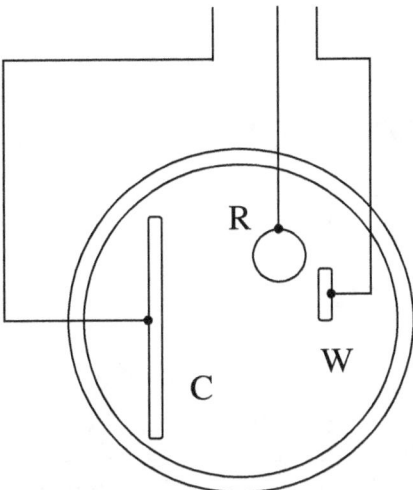

Figure 4.4: Schematic topview of electrochemical cell made from glass. W = working electrode (sample), R = reference electrode, C = counter electrode.

exposed to the electrolyte.
Cyclic Voltametry and Electrochemical Impedance Spectroscopy were the electrochemical techniques to study the GC samples, using a Potentiostat/Galvanostat (EG&G, Model 273A) and an Impedance Gain/Phase Analyzer (Solartron Instruments SI 1260).

4.6 Nitrogen Gas Adsorption

Gas adsorption measurements were carried out [6] on an ASAP 2000 test station from Micromeritics. Nitrogen was used as an adsorbent.
Prior to the measurement, the samples were conditioned by heating and flushing free of contaminating gases and vapor at 200°C.
Isothermal adsorption and desorption plots were recorded, and from single point measurements the BET surface area [49, 50] of the samples was determined.
For powder samples the BET surface area was related to the sample mass in m^2/g.
For monolithic samples with active film and unreacted core material, the overall surface area (the *uptake* in m^2) was related to the geometrical surface of the sample, so that m^2/cm^2 were obtained.
It is remarked that the BET area should not be regarded as the true surface area of a microporous solid [51, 52, 53], because condensation effects can yield an overestimation of the internal surface area.

---

[6] All BET measurements were carried out by Friederike Geiger, PSI.

Chapter 5

Data Analysis

5.1 X-ray Diffraction

Prior to presenting the results of the X-ray diffraction measurements, it is emphasized that numerical values determined here for glassy carbon by means of crystallographical analysis should not be taken absolutely serious, because analysis methods such as baseline (background) subtraction and shape-fitting of the recorded diffractograms were omitted here. Additionally, it is questionable whether X-ray crystallography at all is an appropriate method to fully analyze a material such as glassy carbon, which is a highly disordered material, although much effort was made to study systems which are not represented by simple crystal structures [54, 55, 56]. Rather, GC is more characterized by its disorder than by its order.
Nevertheless, by comparing GC samples which have received different treatment either prior to activation or after activation, i.e. during pyrolysis, one is able to extract qualitative valuable information on structural differences.
One single experimental technique is often not sufficient to characterize the structure of a material such as GC, and other methods to resolve further structural properties are also applied and all results have to be reviewed and brought into a context of understanding.

GC is built up from stacks of graphene layers with a hexagonal structure.
The lateral extension L_a of the graphene layers, their spacing d_{002} and the stack height L_c depend on the HTT [1]. The graphene layers become more extended, their spacing closer, and the graphene stacks higher with increasing HTT [16].
Figures 5.1 and 5.2 show the (002) peaks of G-type and K-type glassy carbon (GC), obtained from samples with 1 mm thickness. From the position of the

---
[1] Heat treatment temperature (HTT) and pyrolysis temperature are termed synonymously in this work

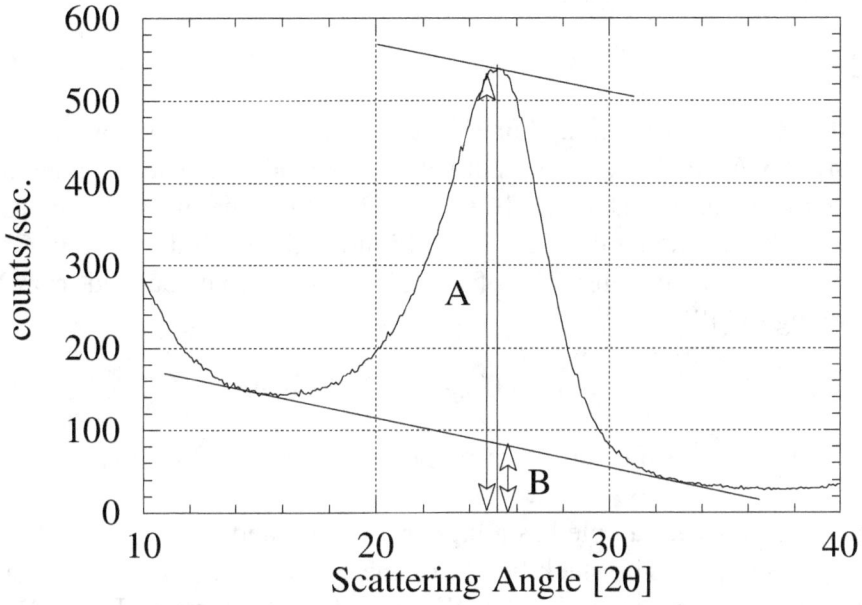

Figure 5.1: (002) peak and R-ratio of G-type GC. The sample thickness was one millimeter. R = A/B = 6.3.

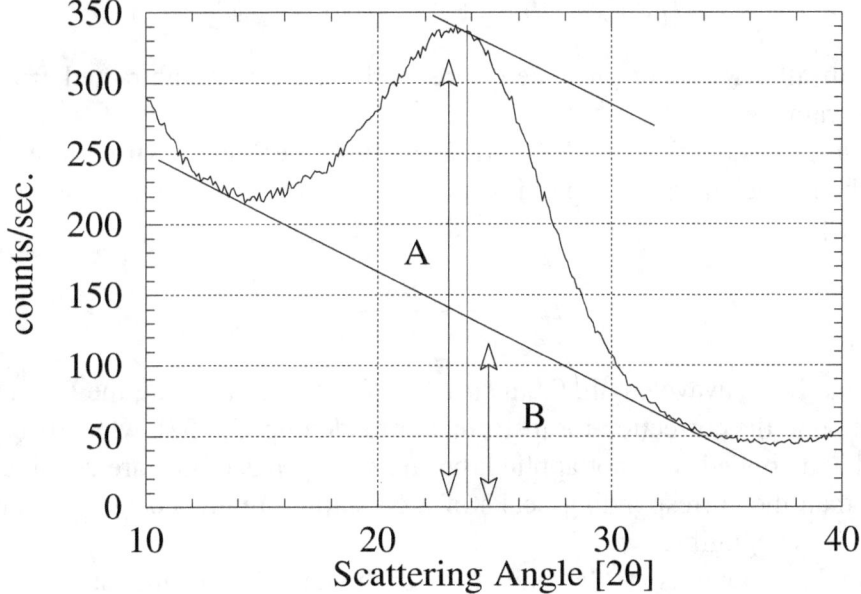

Figure 5.2: (002) peak and R-ratio of K-type GC. The sample thickness was one millimeter. R = A/B = 2.5.

5.1. X-RAY DIFFRACTION

(002) peak (Θ_{002}) in the X-ray diffractogramm, the mean layer distance d_{002} was determined using Bragg's equation:

$$\lambda = 2 \cdot d_{002} \cdot sin\Theta_{002} \qquad (5.1)$$

λ being the X-ray wavelength. For the K-type and the G-type GC a mean d-spacing of 3.68 Å and 3.54 Å, respectively, was obtained. These values do not match with the values in Table 5.1. But they match the general trend that GC with a higher HTT has a smaller d-spacing (see Figure 6.51 and Table 6.10 in Chapter 6, Section 5). The X-ray density ρ_x of the GC sample was calculated from X-ray data according to [48]

$$\rho_x = (3.33538/d_{002}) \times 2.268 \; g/cm^3$$

, where 2.268 g/cm³ is the density of graphite with d_{002}=3.33538 Å.

The peak of the G-type sample has a higher intensity and it is narrower than that of the K-type sample. The stack height of the graphene sheets can be obtained by determining and analyzing the Full-Width-at-Half-Maximum (FWHM) of the (002) peak. The peak width was measured at the intensity

$$I_{FWHM} = I_{back} + 0.5 \cdot (I_{002} - I_{back}) \qquad (5.2)$$

because in all XRD data presented here no baseline was subtracted from the diffractogram.

For further evaluation, the FWHM of the (002) peak can be used to estimate the graphene stack height by applying the Scherrer formula for the crystallite size L_c [57, 55]:

$$L_c = \frac{0.9\lambda}{B \cdot cos\Theta_{002}} \qquad (5.3)$$

λ being the X-ray wavelength, 0.9 a structure factor and B being the FWHM. In this treatment, the correction for instrument broadening [55, 57], which is usually required and applied, was not applied, because the peaks of GC are considerably broader than the corresponding peaks of a crystalline reference sample, and the error made is negligible.

The formula also works for the estimation of the lateral crystallite size L_a, when the FWHM of the (100) peak or (110) peak are known. A structure factor of 0.9 is commonly used. If only two-dimensional lattice reflections of the type ($hk0$) are observed, a structure factor of 1.84 was shown to be correct for the determination of L_a [55].

L_c and L_a can also be determined with Transmission Electron Microscopy (TEM),

HTT [°C]	d [Å]	a [Å]	L_c(002) [Å]	L_a(100) [Å]	L_a(110) [Å]
500	4.8	-	12	-	-
700	4.11	-	13	-	-
900	3.88	-	14	27	-
1000	3.89	2.41	14	29	-
1250	3.9	2.41	15	31	-
1500	3.88	2.41	15	35	35
1750	3.75	2.42	16	38	35
2000	3.58	2.43	22	53	39
2500	3.56	2.43	24	54	48
2700	3.49	2.43	30	60	49

Table 5.1: Variation of crystal parameters in GC with heat treatment temperature (HTT), by *Short* and *Walker* [16, 58]. The crystallite size given by L_a and L_c increases with increasing HTT, while the d-spacing decreases.

and no significant discrepancies in L_c as determined with XRD and TEM are reported in literature. However, the values for L_a determined with TEM deviate by a factor of 10 from the values as estimated with XRD.
Therefore absolute values presented here should not be taken too serious [16].

The effect of pyrolysis temperature on the crystal parameters of GC was studied by Short and Walker [58]. There is a general tendency that the crystallite sizes L_c and L_a increase during pyrolysis, when the temperature is raised. The lattice spacing d decreases with increasing HTT [16]. Data are displayed in Table 5.5.

A comparison of the FWHM of the XRD diffractogramms of G-type (5°) and K-type (7.5°) samples in Figures 5.1 and 5.2 reveals that the G-type sample has crystallites which are thicker and which are laterally more extended than the K-type samples [57].
Depending on the stacking order of the graphene sheets in graphite, two phases of graphite are differentiated:
Phase 2H (JCPDS 26-1079) is a hexagonal phase with stacking order ABAB... , while phase 3R (JCPDS 41-1487) is a rhombohedral phase with stacking order ABCABC.... The two phases are interchangeable by grinding (2H→3R) and by heating (3R→2H) to higher temperatures [59].
Comparing peak positions of K-type and G-type GC, one finds that K-type samples are inclined to the phase 2H and that G-type samples are inclined to phase 3R. This is particularly obvious for the (002) peak with 2H or the (003) peak for 3R, respectively, and for the (101) peak.

5.1. X-RAY DIFFRACTION

In principle it is possible to attain substantially more information about the arrangement and configuration of the graphene layers. The system under investigation can be modelled and the diffraction on the parallel graphene layers can be simulated [56], for instance, by calculating the profile of the (002) peak of GC:

$$I_{002}(q) = f^2(q)\frac{1}{q^2}\left[1 + 2\Re\left(\sum_{n=1}^{M-1}\frac{M-n}{M}\exp^{inq}d_{002}\right)\right], \quad (5.4)$$

$f(q)$ being the atomic scattering factor of carbon, q the scattering vector $4\pi sin\Theta/\lambda$, and M the number of parallel graphene layers with the distance d_{002} from eachother. $\Re$ denotes the real part of the complex expression in parentheses.

In this thesis, attempt was made to calculate I_{002} with *Maple* [2], however, fitting the measured XRD data was not satisfactory, and therefore results are not presented here.

For the quantitative evaluation an empirical value R serves the question, whether the graphitic domains are built up by only one or by more graphene layers [41]. From the ratio of intensities of the (002)- or (003)-peaks [3] I_{002} to the signal background I_{back}, which exhibits an exponential decay, the empirical value R is obtained:

$$R = \frac{I_{002}}{I_{back}}. \quad (5.5)$$

When only single graphene sheets form the carbon, no (002) peak at all is obtained, and the ratio is $R = 1$. For small R only a small portion of parallel ordered graphene sheets exists. If R is measured to be larger, also an accordingly high amount of parallel arranged graphene sheets are present.

If $R = 2$, only 30% of the ordered material consists of single graphene sheets. 70% are built up from parallel arranged (2, 3 or more) graphene layers. Figures 5.1 and 5.2 display the (002) peaks of G-type and K-type glassy carbon samples. The peak height is denoted A in the Figures; the background is denoted B. The K-samples with 1 mm thickness have a signal-to-background-ratio of $R = 2.5$, and the G-samples with 1 mm thickness have $R = 6.3$. Therefore, the graphitic domains in K-samples are to less than 30% (in correspondence to an R ratio larger 2) built up from single graphene layers. In the case of G-type, even less than 30% are built up from single layers [41].

Finally, analysis of the peak height of the (002) peak and (100) peak reveals that the GC samples have an anisotropy in (100) direction [4], because the ratio of the

---

[2] Maple V is a comprehensive computer system for advanced mathematics. It includes facilities for interactive algebra, calculus, discrete mathematics, graphics, numerical computation and many other areas of mathematics. Maple and Maple V are registered trademarks of Maple Waterloo Inc.

[3] 2H: (002) being found at 26.35°; 3R: (003) being found at 26.6°.

[4] private communication with Prof. Hans Grimmer, LNS and PSI

peak heights does not match the ratio given in the XRD reference data (see chapter Experimental). The ratio $I_{002}/(I_{100}+I_{110})$ should be between 5 and 15. However, the ratios are 3.3 for the G-type GC and between 1.5 and 1.9 for the K-type GC samples.

Additional evidence for an anisotropy of the GC samples under investigation was found with SAXS (section 5.6). However, the anisotropy was not further investigated in this thesis.

5.2 Small Angle Scattering

Porous media, such as Glassy Carbon (GC), can be characterized with Small Angle Scattering (SAS). SAS can be carried out either with X-rays or with neutrons. The following considerations and formulae are written for SAXS. However, they can be transferred to SANS as well.

Scattering curves were obtained by radial averaging of the two-dimensional scattering patterns recorded on position sensitive detector plates.

Figure 6.49 displays eight two-dimensional SAXS patterns obtained from two GC samples with 60 microns nominal thickness.

Patterns on the left side were measured in the long geometry (small Q-values) and right patterns were measured in the short geometry (large Q-values).

The four upper patterns (1 - 4) were measured with the sample surface normal parallel to the beam, while the four patterns below (5 - 8) were measured with an angle of 45° between sample surface normal and beam direction.

Patterns 1, 2, 5 and 6 concern the non-activated sample. the other patterns concern the activated (3 hours activated at 450°C) sample.

The patterns from measurements with no tilt-angle are show radial symmetry, while the patterns measured with a tilt show a more complex symmetry, which arises from the fact that now two different length scales are resolved by SAXS.

The graphene sheets probably possess a laminar or a columnar structure. Only further analysis of the patterns allows to distinguish this.

GC is an excellent small angle scatterer [5] and has some remarkable characteristics in its scattering curves. Figure 5.4 displays a typical GC scattering curve in a log-log plot.

Three regimes can be distinguished:

1. The scattering arising from the micropores [60] leads to a steep increasing intensity for scattering vectors $1 \geq Q \geq 0.2$.

2. For $0.05 \leq Q \leq 0.2$ a plateau is observed.

---

[5]GC is often used as a reference material in SAS.

5.2. SMALL ANGLE SCATTERING

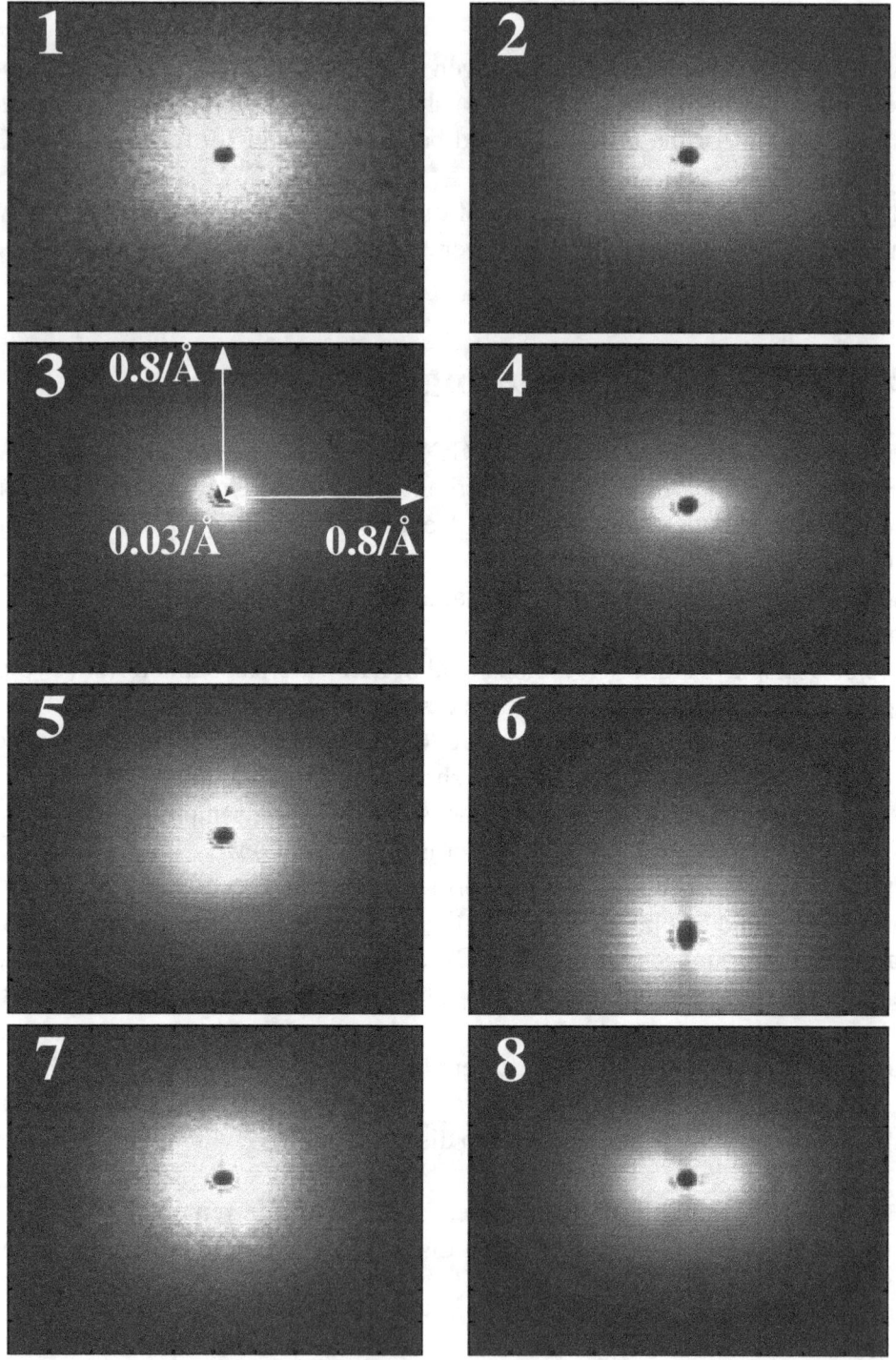

Figure 5.3: 2-dimensional SAXS patterns of GC for short geometry ($0.03/\text{Å} \leq Q \leq 0.8/\text{Å}$). Pictures on left side were measured with the beam perpendicular to the sample surface normal. Pictures on right side were measured with a tilt angle of 45° between sample surface normal and beam direction. Figures 1 and 2 denote a non-activated sample. Figures 3 and 4, 5 and 6, 7 and 8 concern samples activated 30, 150 and 180 minutes at 450°C.

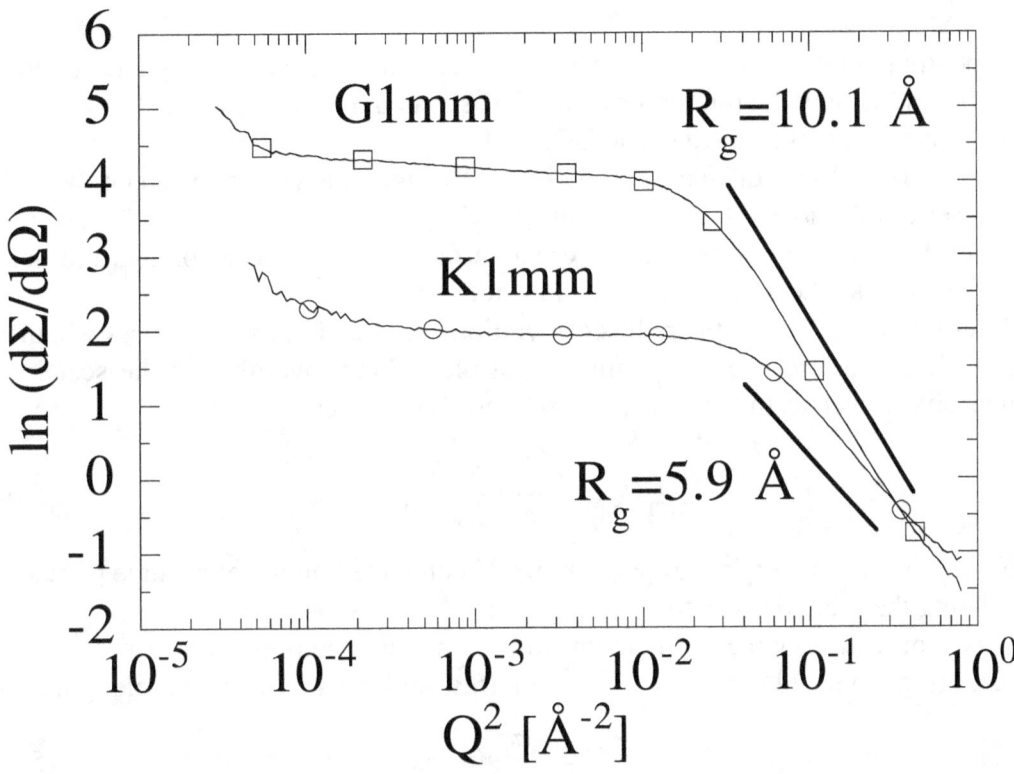

Figure 5.4: Guinier-plot of non-activated GC (K-type and G-type, 1 mm thickness). The radius of gyration R_g was determined from the slope of the curve for $Q^2 \geq 8 \times 10^{-2}$.

5.2. SMALL ANGLE SCATTERING

3. For $Q \leq 0.05$ again a steep increase of intensity is observed, which results from the scattering of pores larger in size than micropores.

The upper resolution limit for SAXS is around 200 Å. Data points with Q smaller than around $8 \cdot 10^{-3}$ 1/Å could not be measured. Beyond this limit only Ultra Small Angle X-ray Scattering can be applied, or Light Scattering. The maximum size which can be resolved with SAXS can be estimated by the reciprocal of $8 \cdot 10^{-3}$ 1/Å, which yields 125 Å. As the resolution limit is given by half of the wavelength, one finds that objects larger than 250 Å can not be resolved anymore by SAXS.

Depending on the sort of GC investigated, the scattering curves may alter a little. In this thesis only the micropores were investigated, because they contribute most to the internal surface area of the GC.

Similar to the X-ray diffraction experiments, various sorts of untreated GC as well as a series of oxidized GC were investigated.

From the scattering curves, the internal surface area, the pore volume, the pore shape and also the pore radii can be inferred.

After Guinier [61, 62], the radius of gyration R_g of an object, regardless whether particle or void, can be determined by a plot of the logarithm of the scattered intensity $\frac{d\sigma}{d\Omega}$ versus the square of the scattering vector Q:

$$\frac{d\sigma}{d\Omega}(Q) = \frac{\Delta n_f^2 NV^2}{exp\left((QR_g)^2/3\right)}, \tag{5.6}$$

N being the number of pores per volume, V being the volume of a single pore and Q being the scattering vector.

Q is a measure for the momentum transfer of the electrons with respect to the scattering atom and is related to the scattering angle Θ and X-ray wavelength λ as

$$Q = \frac{4\pi}{\lambda} sin\,\Theta\,. \tag{5.7}$$

The scattering contrast Δn_f^2 is the difference of the electron densities in the vicinity of a particle (or void) in a matrix and depends on the chemical composition and mass density:

$$\Delta n_f^2 = \left(\sum_i f_i(n_{i,Part.} - n_{i,Matr.})\right)^2, \tag{5.8}$$

f_i being the atomic structure factor [47] of atoms of sort i and n_i being the number density of atoms of sort i. For $Q = 0$, the measured intensity is

$$\frac{d\sigma}{d\Omega}(Q=0) = \Delta n_f^2 NV^2\,, \tag{5.9}$$

in literature being referred to *scattering at zero angle* or *forward scattering*. This quantity is proportional to the number of voids and to the square of the volume of a single void and thus yields valuable information of the system studied.

The radius of gyration is defined as the moment of the electron density distribution and is being written as a formula well known by the inertia momentum in classical mechanics:

$$R_g^2 = \frac{\int_V r^2 \Delta n_f(\vec{r}) d^3\vec{r}}{\int_V \Delta n_f(\vec{r}) d^3\vec{r}}, \qquad (5.10)$$

$\rho(r)$ being the electron density [63]. The radius of gyration is not the geometrical radius of the object under investigation (pores or voids). However, there are formulae (Table 5.2 and Ref. [64]) to convert R_g into the actual geometrical dimensions as in the case of spheres, cylinders, ellipsoids or cubes. For spheres, following relation between R_g and the sphere radius R holds [64]:

$$R_g = \sqrt{\frac{3}{5}} R \qquad (5.11)$$

for other particle shapes a more general expression is known:

$$R_g = \sqrt{\frac{n_1 + c^2}{n_2}} R, \qquad (5.12)$$

the geometry factors of which are listed in Table 5.2.

The advantage of determining the radius of gyration is that no specific geometry

Shape	n_1	n_2	c
Sphere	2	5	1
Cube	2	3	1
Disc	1	4	1
Needle	1	12	0

Table 5.2: Geometry factors for the conversion of the gyration radius R_g into the actual radius R (after [64]).

of the objects under investigation has to be assumed. The radius of gyration can be obtained from experimental data either by fitting data points to equation 5.12 or by evaluating a so-called Guinier-plot, when the logarithm of the scattered intensity is plotted versus the square of the scattering vector [65].

5.2. SMALL ANGLE SCATTERING

The range with a straight line in this plot is called Guinier range. The slope of the straight line equals R_g^2. Equation 5.12 is, strictly spoken, only valid for randomly distributed voids in a diluted system [66, 67, 68, 69].

When voids or particles are densely packed, equation 5.12 must be corrected by an interference function. However, in the case of a not too sharp pore size distribution, as in the case of carbons, this is omitted [30].

If the voids are not randomly distributed, for instance, when an anisotropy is present, equation 5.12 is also corrected by an anisotropy function.

There is experimental evidence that the GC samples have an anisotropy. Firstly, the XRD data reveal a presence of an anisotropy, because the peak height of the (002) peak is larger than the (001)-peak or (010)-peak, as would be expected from the XRD reference patterns.

Secondly, SAXS curves were recorded from GC samples also with a 45° tilt angle versus the surface normal of the samples. No rotation symmetric scattering pattern was obtained. Instead, the scattering pattern was laterally distorted, giving rise to the suggestion that the samples have an anisotropy. Due to the limited beamtime it was not possible to gather sufficient data for an anisotropy function.

Scattering curves can also be modelled by calculating an explicit shape factor of the objects. The structure of the objects has to be known or at least assumed. The shape factor is defined as

$$S_1(\vec{Q}) = \left| \frac{1}{V} \int_V e^{i\vec{Q}\vec{r}} d\vec{r} \right|^2 \qquad (5.13)$$

For $Q = 0$, the shape factor becomes unity: $S_1(0)=1$.

Shape factors are known for many differently shaped particles and listed [64]. For special purposes shape factors can be calculated even if the objects lack in high symmetry.

For the quantitative analysis of our scattering data, it was assumed that the micropores of the GC are spherelike and their size distribution is a logarithmic normal distribution [65]. Among the many functional forms of a pore radii distribution applied for fitting experimental data (Weibull distribution, Γ distribution and various others), the log normal distribution has the advantage that its product with a power of the radius is analytically integrable [70].

In addition, it was neglected that the samples could have an anisotropy. For our investigations we used the shape factor S(Q,R) for a sphere with radius R, which is given by [71]

$$S(Q, R) = \left(3 \frac{sinQR - QRcosQR}{(QR)^3} \right)^2 . \qquad (5.14)$$

This equation yields a plot with oscillations, which occur at values for QR when the denominator becomes zero. In none of our samples investigated with SAS such oscillations occurred. Therefore it was concluded that the pore radii distribution was not sharp, but rather broad. This result also served as a justification that there was no necessity to correct Equation 5.12 by an interference function.
However, as it was obvious that the radii were not sharply distributed in the system, a pore radii distribution had to be applied for further data analysis.
The following logarithmic normal distribution P(R) was chosen:

$$P(R) = \frac{1}{\sqrt{2\pi}} \cdot \frac{1}{R \cdot \sigma} \cdot exp\left(-\frac{ln^2 \frac{R}{R_0}}{2\sigma^2}\right), \int_{R=0}^{\infty} P(R)dR = 1 \qquad (5.15)$$

where R_0 = radius in the maximum of the distribution, σ = width of the distribution. The only justification for choosing a logarithmic normal distribution is that the mathematical treatment of the data analysis is facilitated by it. There is no physical justification, but this treatment is widely accepted. The scattering cross section can then be written as [67]:

$$\frac{d\sigma}{d\Omega}(Q) = \Delta n_f^2 \cdot N \cdot \int_{R=0}^{\infty} V^2(R) \cdot S(Q,R) \cdot P(R) \cdot dR \qquad (5.16)$$

where N denotes the number of pores, $V=\frac{4}{3}\pi R^3$ is the volume of a single pore, and Δn_f is the scattering contrast of GC, which is evaluated in the experimental part of this work [72].
Using Equation 5.16, the scattering curves of the activated GC samples, corrected as described in the experimental part, were fitted. Figure 6.56 (Chapter 6, Section 5) displays a SAXS curve of a K-type GC sample with 55 microns thickness (not activated). The drawn line is the best fit function as obtained by applying equation 5.16.
The scattering from the mesopores (measured at small Q-values) was neglected in so far as the scattering intensity arising from micropores was extrapolated to small scattering vectors.
The data points in Figure 6.56 with $Q \leq 0.05$ arise from the scattering of the mesopores and are above the fitted curve. The intercept of the fitted curve with the abscissa is the micropore forward scattering or scattering at zero angle, respectively.
The extrapolated value was taken as the micropore scattering at zero angle. [6] In the fitting procedure, three independent parameters were varied: the width σ of the distribution, the radius R_0 in the maximum of the distribution, and the scattering at zero angle.

---

[6]For further treatment of scattering data, the micropore scattering contribution (fitted curve) must be subtracted from the scattering curve [73].

5.2. SMALL ANGLE SCATTERING

Using the two former parameters, the pore size distribution can be plotted, as displayed in Figure 6.57.
The scattering intensity at zero angle for the micropores is given by Equation 5.17:

$$\frac{d\sigma}{d\Omega}(Q \to 0) = \Delta n_f^2 \cdot N \cdot \langle V^2 \rangle \qquad (5.17)$$

where $\langle V^2 \rangle$ is the mean square of the micropore volume:

$$\langle V^2 \rangle = \int_{R=0}^{\infty} \left(\frac{4}{3}\pi R^3\right)^2 \cdot P(R) \cdot dR. \qquad (5.18)$$

Values of the mean micropore radii $\langle R \rangle$ and their corresponding mean surface area $\langle A \rangle$ can be calculated according to

$$\langle R \rangle = \int_{R=0}^{\infty} R \cdot P(R) \cdot dR \qquad (5.19)$$

and

$$\langle A \rangle = \int_{R=0}^{\infty} 4\pi R^2 \cdot P(R) \cdot dR. \qquad (5.20)$$

The total pore volume of the pores can be calculated using the invariant Q_0, which is obtained by integrating the scattering curve over the whole Q-range [67, 66]. No specific pore model is necessary to obtain this information:

$$Q_0 = \int_{R=0}^{\infty} \frac{d\sigma}{d\Omega}(\underline{Q})d^3Q = (2\pi)^3 \cdot \Delta n_f^2 \cdot N \cdot V \qquad (5.21)$$

The porosity p or void fraction v ($p = 100\,v$) can be calculated according to

$$v = N \cdot V/V_{sample}. \qquad (5.22)$$

After Porod [68, 69, 66], the internal surface area $N \cdot A$ can be determined from a Porod plot of IQ^4 vs. Q^4. Again, no specific pore model has to be assumed:

$$\frac{d\sigma}{d\Omega}(QR \geq 5) = \Delta n_f^2 \cdot 2\pi \cdot N \cdot A \cdot Q^{-4} \qquad (5.23)$$

The volumetric surface area s was determined according to

$$s = N \cdot A/V_{sample}. \qquad (5.24)$$

Strictly spoken, Equation 5.23 is only valid for isometric and isotropic particles with a smooth surface. Such particles have an asymptotic scattering behaviour such that the slope in a log-log plot is proportional to Q^{-4}. In many cases this behaviour is not observed. Instead, the exponent n of decay is smaller than 4. Insufficient collimation of the X-ray beam is one reason for this effect.

The problem of insufficient collimation arises always when a conventional X-ray source (cathode) is applied. Special desmearing procedures have to be applied in raw data analysis for correction of this effect. [74, 75].

When synchrotron radiation is applied, the X-ray beam is well collimated and no desmearing procedures are necessary

Prior to the application of synchrotron radiation, GC K-type and G-type samples with 1 mm thickness were measured with the Philips X-ray Diffractometer. The exponent n of decay of the intensity for the small Q-values (mesopores) in a log-log plot was not very close to 4 for both samples: for the K-type it was n = 1.9 and for the G-type it was n = 3.1.

Another reason for a deviation from Porod's law is that the surfaces are not smooth, but rough. Peterlik et al. [76] found that the exponent of Q is increasing for GC when the HTT is raised. Therefore one may conclude that the internal surfaces of the GC mentioned above are not smooth, but rough [77].

Nevertheless it is widely accepted to determine the Porod constant from the scattering data after a subtraction of a constant, so that the scattering curve is forced to show a behaviour with a decay of 4. This procedure is displayed in Figure 5.5:

Finally, from the slopes of the scattering curves in a log-log plot the geometry of objects (particles or pores) can be inferred [66]. Smooth and isotropic particles have a decay with n = 4. The asymptotic form of the power law for the decay of scattered intensity arising from objects with specific geometry is listed in Table 5.3.

Sometimes fractal structures occur, i.e. in stochastic growth processes [77, 66].

Scatterer	Exponent n of decay
Sphere	-4
Thin disk	-2
Thin cylinder	-1
Spherical shells	-2
Random fluctuations	-4
Line dislocations	-3

Table 5.3: Asymptotic behaviour of various scatters for large Q [64].

There are volume fractals and surface fractals. Structure relevant parameters of

5.2. SMALL ANGLE SCATTERING

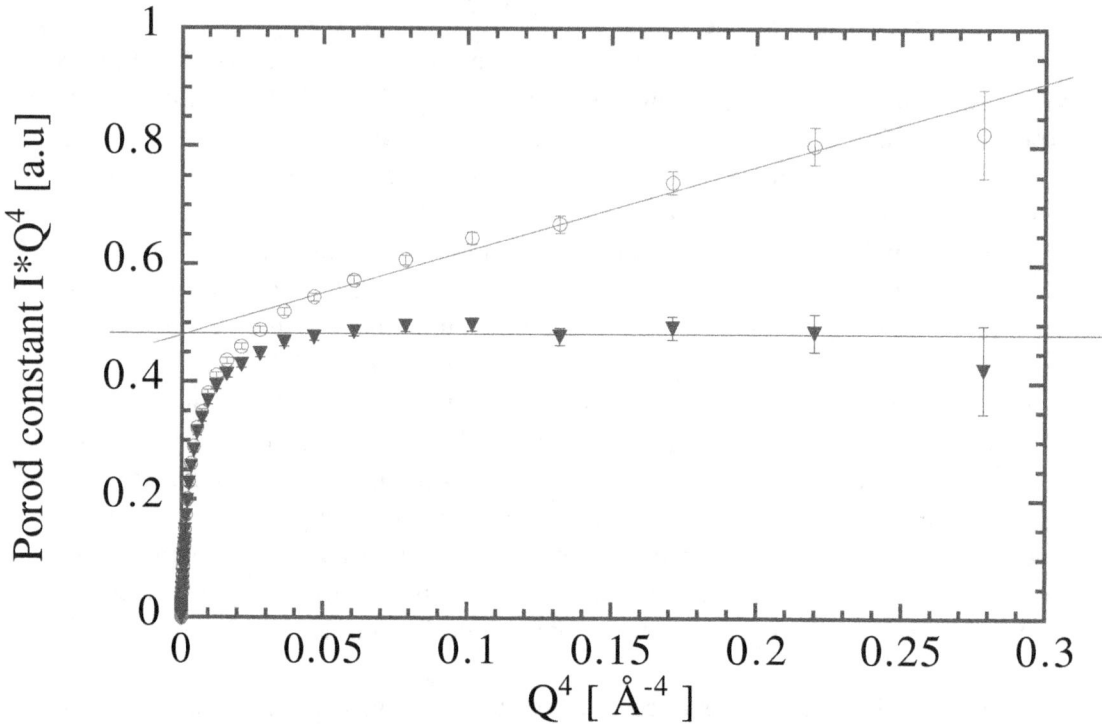

Figure 5.5: Porod-plots of scattering curves from K-type GC with 60 micron nominal thickness, obtained after 3 hours of activation at 450°C. Open circles concern the raw data curve. As this curve has a slope $\neq 0$ for the large Q values in the Porod plot, it is clear that the Porod law of Q^{-4} in Equation 5.23 is not recovered. The triangles denote the curve with a constant background subtraction. The constant was chosen so that the slope for large Q values becomes zero. The straight lines are fitted curves for the values $QR > 5$. The intercept with the abscissa is the Porod constant. Note, that the Porod constant is same for raw data curves and for corrected curves.

objects with fractal structure are their fractal dimensions D.
Porous objects have fractal dimensions of $1<D<3$ and are called volume fractals. The volume V of a cluster shows a scaling behaviour like

$$V \sim R^D, \qquad (5.25)$$

R being the radius and D being the fractal dimension, $1<D<3$. The limiting case $D = 1$ describes the non-physical case of a cylinder whose length is infinite. D=3 is the limiting case with the euclidean dimension of a compact and homogeneous particle, a volume fractal with a smooth surface. The scattered intensity then yields

$$\frac{d\sigma}{d\Omega}(Q) \sim Q^{-D}, \text{ and } 1 \leq D < 3 \qquad (5.26)$$

For the limiting case $D = 3$ above equation is not valid anymore. Rather the Porod behaviour for an assumed smooth surface in such cases [66] is recovered.

Rough objects are characterized by a decay of D_s-6 with $3 \leq D_s < 4$ and are called surface fractals. D_s is the dimension of the fractal surface. When the surface is extended throughout the whole particle volume, the limiting case $D_s = 3$ is obtained. Surface fractals are compact objects (D=3) with a rough surface, the area A of which exhibits a scaling behaviour like

$$A \sim R^{D_s} \qquad (5.27)$$

and $2<D_s<3$; R being the radius of the object. The case $D_s = 2$ is recovered for a smooth surface and represents a limit (Porod). D_s=3 is an extreme case, at which the surface is extended throughout the whole particle volume. The scattered intensity for large Q is [66]

$$\frac{d\sigma}{d\Omega}(Q) \sim Q^{D_s-6}, \text{ and } 2 < D_s < 3 \qquad (5.28)$$

For fractal structures different, also *fractal* exponents, may occur, as listed in Table 5.4:

5.2. SMALL ANGLE SCATTERING

Objects	Exponent n
Porous objects (volume fractals)	1 ... 3
Rough particles (surface fractals)	3 ... 4
Smooth particles, isometric	4 (Porod)

Table 5.4: Depending on the *fractality* of objects, different scaling behaviour with according exponents n is observed [64].

5.3 Electrochemical Characterization

5.3.1 Cyclic Voltametry

Cyclic Voltametry (CV) [7] was applied to check whether electrochemical reactions occur within the potential range from 0.0 to 1.0 Volt. Additionally, CV was used to check the electrochemical system for cleanliness.
In CV, an alternating linearly increasing and decreasing dc voltage is applied to the electrode, and the current response is measured.
The potential transient at the working electrode is displayed in Figure 5.6.
The experiment is starting at an initial potential $\phi_{u'}$ with a constant scan rate ν:

$$\nu = \frac{d\phi}{dt} = \frac{\phi_{u''} - \phi_{u'}}{\Delta t}. \quad (5.29)$$

If at some particular potentials specific reactions occur, a current evolves which can be measured.

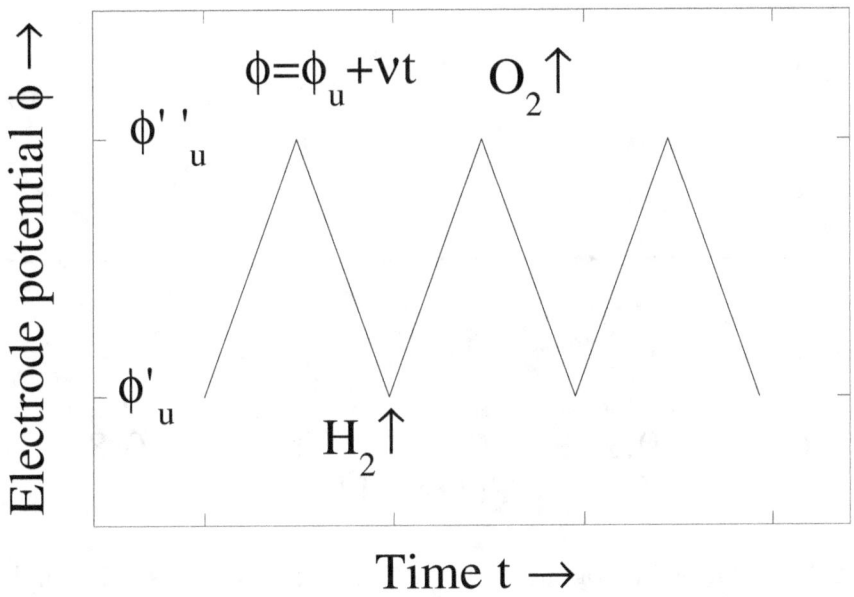

Figure 5.6: Potential/time relationship at a working electrode during cyclic voltammetry. Negative (ϕ'_u) and positive (ϕ''_u) reversal potential are usually chosen between the evolution of hydrogen and oxygen.

Figure 5.7 displays a cyclic voltamogram (CV) of a GC sample, activated 96 minutes at 450°C in air and measured in 3 molar sulfuric acid as electrolyte between

5.3. ELECTROCHEMICAL CHARACTERIZATION

0.00 and 1.00 Volt. The shape of the CV is typical for activated GC measured in sulfuric acid.

The current density of around 2 mA/cm^2 between 0.00 and 1.00 Volt arises mainly from the double layer charging of the activated GC surface.

Around 0.40 Volt (extending from 0.2 to 0.65 Volt) there is a broad current wave indicating a redox process (or even several processes) taking place in the system. This current wave is attributed to the so called quinone/hydroquinone redox couple on activated GC [78, 79, 80, 81, 82, 83].

From 0.7 Volt on the forward scan on, the current increases, possibly due to the creation of an oxygen chemisorption layer on the GC surface, which is removed on the backscan.

The rectangle drawn in the Figure 5.7 describes the CV of an ideal capacitor with

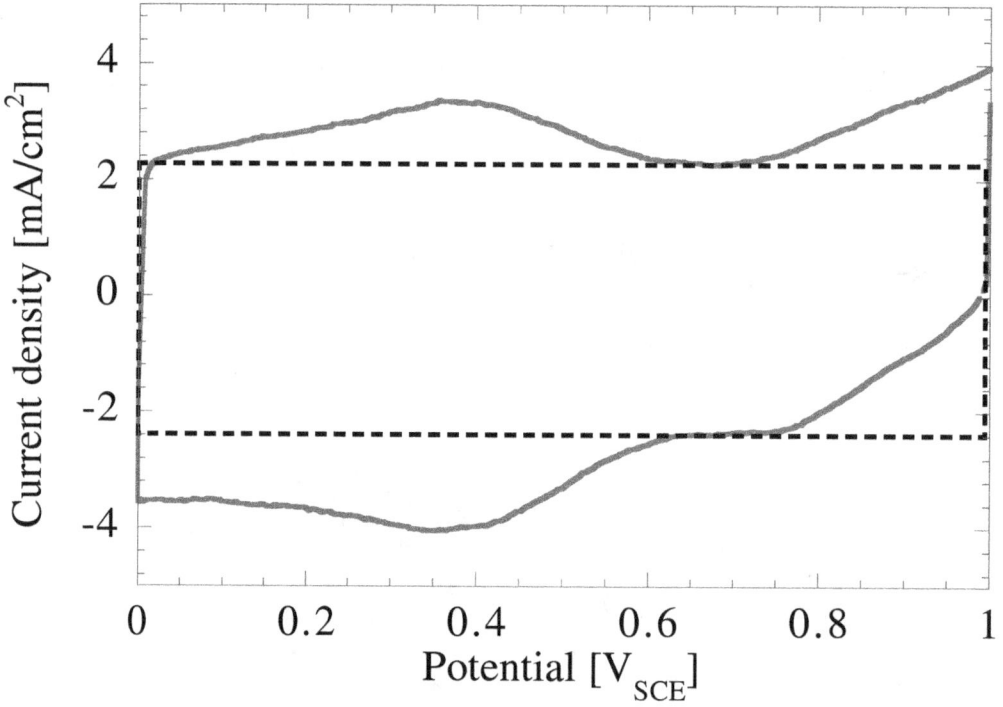

Figure 5.7: Cyclic voltamogram of an activated SIGRADUR®K sample. The scan rate was 10 mV/sec. The dashed line depicts a rectangular CV of an ideal capacitor with only double layer charging and discharging.

the charging and discharging of the double layer.

As a general result one may extract that the capacitance measured in GC depends on the bias voltage applied.

The capacitance can be determined from the current density and the scan rate:

$$C = \frac{dQ}{d\phi} = \frac{jdt}{d\phi} = \frac{j(\phi)}{\nu}, \qquad (5.30)$$

ν being the scan rate in mV/sec and j being the current density in A/cm^2.
Redox processes may enhance the capacitance of oxidized GC by a factor of 1.5 to 3; usually there is a ratio of around 2 between the capacitance measured at 0.4 Volt and 0.9 Volt.
Redox processes can contribute to a large extent to the overall capacitance. However, in the case of GC, this contribution is restricted to the rather low potential of around 0.4 Volt. When a capacitor with two electrodes is considered, these redox processes take place at 0.00 Volt, at which the capacitor is discharged. Therefore, these redox capacities yield no advantage for the capacitor. Additionally, it is well known that redox processes can cause corrosion and degradation of electrode materials.
Especially in the case of carbon electrodes, redox processes should be avoided, because carbon may react in many ways in electrochemical systems with the electrolyte and environment. This can be concluded from the so called Pourbaix diagramm of carbon [15], which states that carbon has no stable potential window in aqueous electrolytes.

5.3.2 Electrochemical Impedance Spectroscpy

The measurement of complex resistances (Impedance Z) with AC voltage techniques (Electrochemical Impedance spectroscopy, EIS) is an important tool to investigate interface and volume properties of materials [84].
When interfaces are studied, such as electrode-electrolyte interfaces, adsorption-rates and reaction-rates can be studied as well as double layer capacitance.
In this thesis, EIS was applied to determine the DLC of Glassy Carbon electrodes. If a system is excited with an ac voltage, say

$$U(\omega) = U_0 \cdot exp(i\omega t), \qquad (5.31)$$

U_0 being the amplitude, f being the frequency and

$$\omega = 2\pi f, \qquad (5.32)$$

then the current $I(\omega)$ flowing through the system in general will be shifted against the voltage by a phase shift ϕ:

$$I(\omega) = I_0 \cdot \exp(i\omega t + \phi), \qquad (5.33)$$

5.3. ELECTROCHEMICAL CHARACTERIZATION

I_0 being the amplitude of the ac current. The impedance follows from Ohms law:

$$Z(\omega) = \frac{U(\omega)}{I(\omega)} = Z_0 \cdot exp(i\phi) = Z_0 cos\phi - iZ_0 sin\phi. \qquad (5.34)$$

A schematic of the impedance, showing the phase shift between current and potential, is illustrated in Figure 5.8.

The real part (imaginary part) of the complex impedance, $Z_0 cos\phi$ ($Z_0 sin\phi$), will

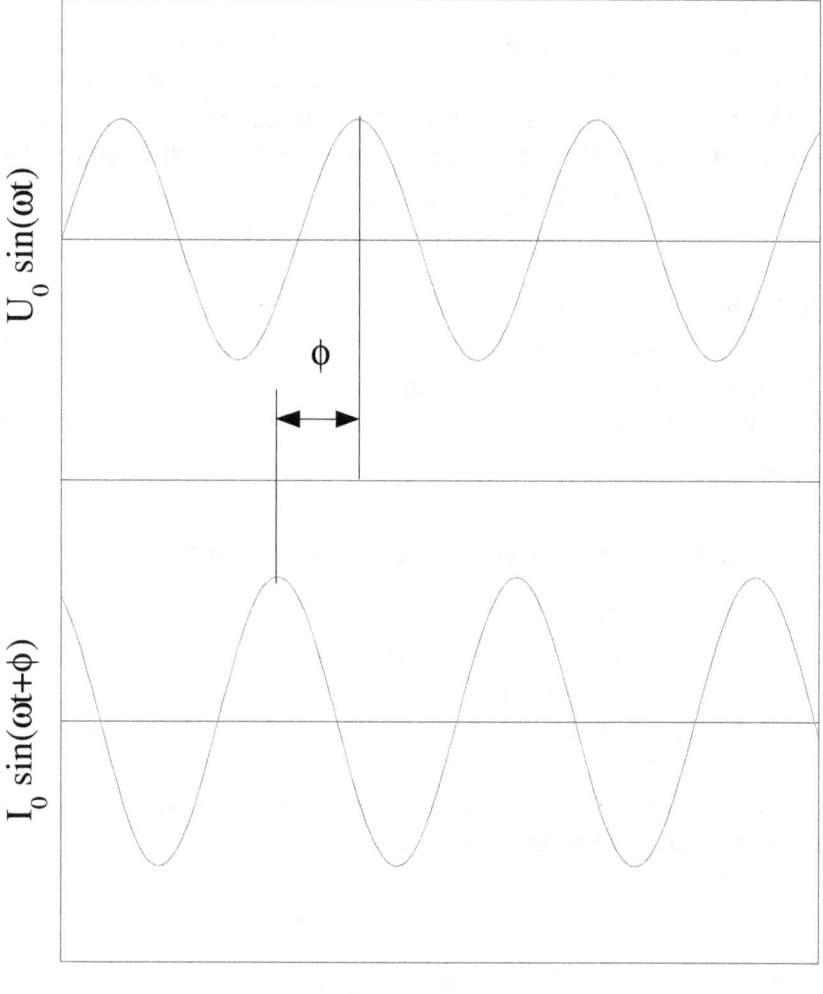

Figure 5.8: Schematic of Impedance. The phases of current and voltage are shifted by the amount of $\Delta\phi$ against eachother: $\Delta\phi = \omega\Delta t$.

be abbreviated with Z' (Z'').

Complex Resistances

Ohmic resistance of a resistor	$Z_R = R$
Capacitance of a capacitor	$Z_C = \frac{1}{i\omega C}$
Inductance of a coil	$Z_L = i\omega L$

Table 5.5: Resistivities of various electrotechnical elements. Coil and capacitor depend on the frequency ω of the applied ac voltage.

In many cases systems can be modelled by an array of resistances R, capacitances C and inductances L.

Table 5.5 displays the frequency response of these simple elements: The impedance of a combination of R, L and C can be calculated using Kirchhoffs rules.
For a series circuit of two impedances Z_1 and Z_2, the impedance is

$$Z_{1,2} = Z_1 + Z_2. \tag{5.35}$$

For a parallel circuit of two impedances Z_1 and Z_2, the impedance is calculated as follows:

$$\frac{1}{Z_{1,2}} = \frac{1}{Z_1} + \frac{1}{Z_2}. \tag{5.36}$$

Representation of Electrochemical Systems by Electric Circuits

Figure 5.9 displays a schematic of an electrochemical interface, which can be represented by the following electric circuit [7]:
In the schematic, R_T denotes the charge transfer resistance, which is attributed to chemical reactions occurring on the electrode surface.
 This resistance is infinite, when no redox processes occur on the electrode.
The electrochemical double layer has a capacitance C_D parallel to R_W.
If chemical reactions occur, the concentration of ions may decrease in front of the electrode with the result that an additional resistance R_W in series with a capacitance C_W also occurs in the system, in series to the charge transfer resistance and parallel to the double layer capacitance [7].
 This element is known as the *Warburg-Impedance* and is characterized by a straight line with an angle of 45° against the real axis in the complex plane:

---

[7] Although termed as *occurrence of a resistance*, the correct speech means the occurrence of a conductivity.

5.3. ELECTROCHEMICAL CHARACTERIZATION

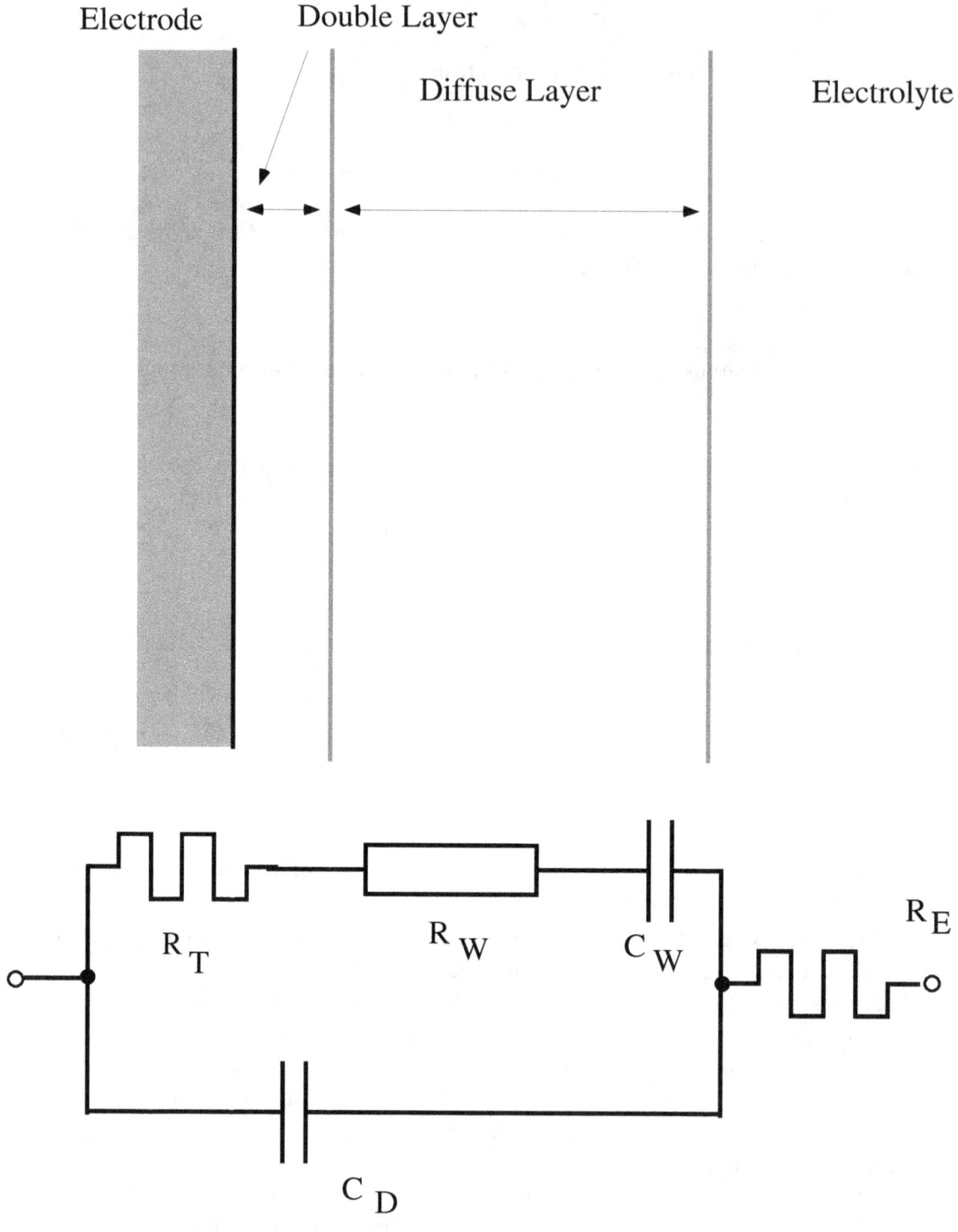

Figure 5.9: Schematic of the electrochemical interface and its corresponding electric circuit.

$$\mathcal{R}_W = R_W + \frac{1}{i\omega C_W} \,. \tag{5.37}$$

Finally, the resistance of the electrolyte, R_E, contributes to the overall resistance of the system in series.

Using the complex resistances in Table 5.5 and Kirchhoff's rules (Eqns. 5.35 and 5.36), one finds for the impedance of the circuit in Figure 5.9:

$$Z = \frac{1}{\dfrac{1}{R_T + R_W + \dfrac{1}{i\omega C_W}} + i\omega C_D} + R_E \,. \tag{5.38}$$

Note that for high frequencies ω the limit of the impedance Z is the electrolyte resistance R_E.

In the case of an electrochemical double layer capacitor electrode, if redox reactions do not take place and if charge transfer resistances are very high, there will also be no remarkable concentration gradients of electrolyte ions.

Then, the situation can be represented by a simple series of a resistance and a capacitance, as displayed in Figure 5.10.

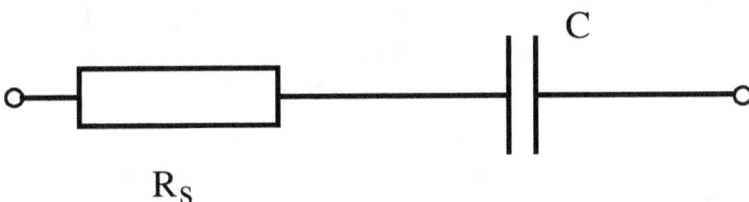

Figure 5.10: Series circuit of a resistance R_s and a capacitance C, representing the most simple capacitor electrode (single electrode).

The impedance of this circuit equals

$$Z = R_s - \frac{i}{\omega C} \,. \tag{5.39}$$

The complex impedance Z can be split in the real part Z' and imaginary part Z":

$$Z = Z' + iZ'' \,. \tag{5.40}$$

5.3. ELECTROCHEMICAL CHARACTERIZATION

For the imaginary part we find

$$Z'' = -\frac{1}{\omega C}. \qquad (5.41)$$

Thus, the double layer capacitance of a system which is represented by the circuit in Figure 5.10 can be determined from the imaginary part of the measured complex impedance:

$$C = -\frac{1}{\omega Z''} \qquad (5.42)$$

All capacitance values for activated GC single electrodes reported in this thesis were determined using equation (5.42).

In real capacitors with two electrodes, often a leakage current at open circuit may occur with the result of a self discharge of the capacitor.
The leakage current is taken into account by an additional parallel resistance R_p,

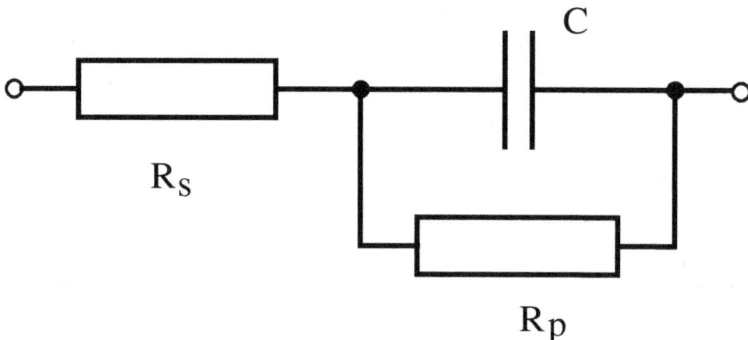

Figure 5.11: Representation of the electric circuit of an EDLC.

as displayed in Figure 5.11.
The real part and imaginary part of the complex impedance of this circuit are

$$Z' = R_s + \frac{1/R_p}{\omega^2 C^2 + 1/R_p^2}, \qquad (5.43)$$

$$Z'' = \frac{-\omega C}{\omega^2 C^2 + 1/R_p^2}. \qquad (5.44)$$

From experimental impedance data all information on R_s, R_p and C can be extracted.

From the high frequency intercept of the impedance plot, the series resistance R_s can be extracted:

$$\lim_{\omega \to \infty} Z' = R_s. \qquad (5.45)$$

The information on R_s is necessary to solve equations (5.44) and (5.45) for the values for **R_p** and **C**:

$$C = -\frac{Z''}{\omega \left(-2 Z' R_s + R_s^2 + Z''^2 + Z'^2\right)}, \qquad (5.46)$$

$$R_p = -\frac{-2 Z' R_s + R_s^2 + Z''^2 + Z'^2}{-Z' + R_s}. \qquad (5.47)$$

Often experimental data deviate from the theoretical relations in so far as a pure capacitive contribution of C_D is not sufficient to describe the true behaviour of the circuit in Figure 5.11.

Then, phenomenologically a so-called *constant phase element* p is introduced:

$$Z_p = \frac{1}{(i\omega)^p C} \qquad (5.48)$$

with the real number $0 \le p \le 1$.
Often p is close to 1.
The expressions for **R_p** and **C** can be calculated in the same way as in the case without p.
However, these expressions are more complex:

$$R_p = \frac{(Z'^2 - 2Z'R_s + R_s^2 + Z''^2)\sin(\chi)}{R_s \sin(\chi) - Z' \sin(\chi) + Z'' \cos(\chi)}, \qquad (5.49)$$

$$C = -\frac{Z''}{(Z'^2 - 2Z'R_s + R_s^2 + Z'^2)\exp(1/2 p \ln(w^2))\sin(\chi)}. \qquad (5.50)$$

with

$$\chi = \frac{1}{2} p \, signum(w) \pi. \qquad (5.51)$$

The capacitance values in this thesis were determined from equation 5.42 with a reasonable accuracy.

5.3.3 Diffusive Resistance of Porous Electrodes

Due to the porous active film the GC electrodes exhibit a typical porous electrode behaviour.

5.3. ELECTROCHEMICAL CHARACTERIZATION

While an ideal capacitor electrode shows a vertical line in the Nyquist representation of the impedance spectra (Z" versus Z'), a porous electrode bends over towards smaller impedance values at very high frequencies [85].
The regime of bending over is indicated ideally by a straight line with an angle of 45° with respect to the real axis of impedance. This behaviour of the impedance can be attributed to the limited diffusion of ions in the pore filled with electrolyte in a high frequency field [86] and is in so far a transport resistance (similar to a so-called *Warburg impedance*).
This resistance is a series resistance, which is schematically assigned as an ohmic series resistance to the capacitive resistance (Figure 5.11). The diffusive resistance of a single porous electrode can be obtained by evaluating the high frequency intercept of the impedance spectra in the Niquist representation.
Figure 5.12 displays an impedance curve of a thermally activated GC sample in Nyquist representation. Measured data points are connected by a solid line. The

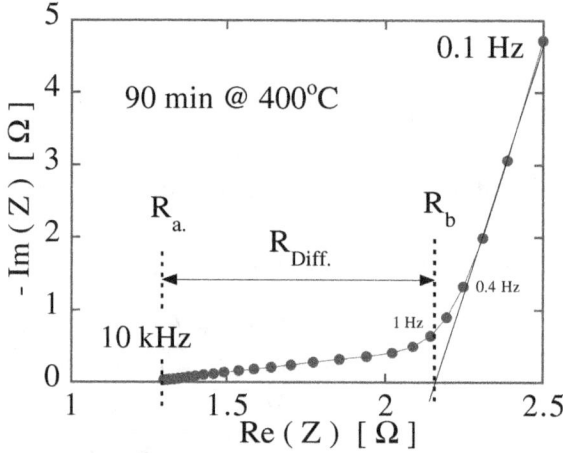

Figure 5.12: Determination of diffusive resistance of porous electrodes. The intercept of the extrapolated tangent to the impedance curve with the real axis towards high frequencies and the high frequency intercept of the actual curve determine the diffusive resistance R_{Diff}.

tangent on this line extrapolating to high frequencies has an intercept with the real axis at position R_b. The actual impedance curve has an intercept at position R_a. The difference $R_b - R_a$ is the diffusive resistance of the electrode, a measure for the electric resistance due to ion transport limitation [85]. This resistance must be related to the electrode surface or even to the active film volume, so that a specific or volumetric diffusive resistance is obtained.

Chapter 6

Results and Discussions

6.1 Thickness and Mass Changes during Activation

The activation of GC is an oxidation process which leads to a thinning of the overall sample and to a decrease of the overall mass. At least two different processes take place during activation:

- The sample is getting thinner so that after a specific time the whole sample is converted to CO, CO_2 and ash.

- A film with open pores is created in the sample so that after a specific time the sample consists of active material only.

Both processes are accompanied by mass lost. Figure 6.1 displays a fracture cross section of a partially activated GC sheet sample.

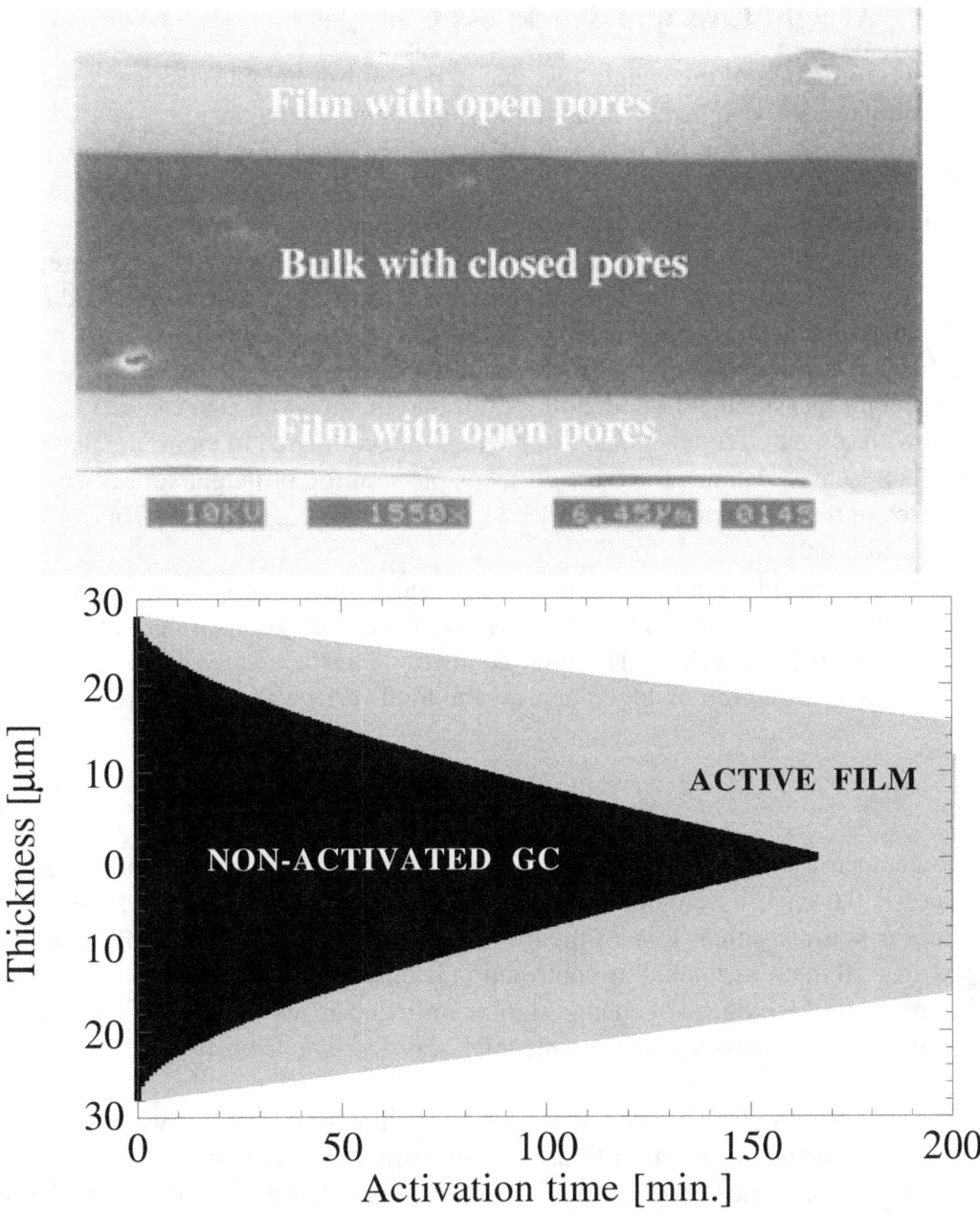

Figure 6.1: Top: Scanning electron micrograph of a partially activated glassy carbon sheet (49 minutes at 450°C). The image represents a fracture cross section of a SIGRADUR®K sample with 60 microns nominal thickness. The unreacted core material appears black and is enveloped by a film with open pores, which appears bright. Bottom: Schematic of the evolution of sample thickness and film thickness during thermochemical activation.

6.1.1 Weight Loss and Thickness Changes

To get information of the weight loss due to burn-off of the GC samples, some of the samples were weighed on a laboratory high precision balance.
The weight loss of the SIGRADUR®K samples with 1 mm thickness after 30 minutes of activation at 550°C was around 10%. After 3 hours of activation the samples were only half as thick as before activation.
The decrease of thickness is particularly important when thin GC sheets are activated, because a long activation can cause the samples to disappear totally due to burn-off. The thickness change of the SIGRADUR®K samples with 100μthickness at 490°C activation temperature was around 10 to 20 μm per hour [87].
The weight measurement yielded the result that the mass change (dm/dt) is constant with activation time. As the samples are thin compared to their lateral extension (1×2 cm² ×1 mm), the burn-off takes place at the principal surfaces of the samples in majority and to a much lesser extent at the edge planes. This will be exercised briefly:
Consider a cuboid sample with thickness c and length a and width b. The geometrical area of the principal planes is $2S_p$=2ab. The geometrical area of the edge planes is S_e=2(a+b)c. The total geometrical surface area of the sample is $S_{tot.}$=$2S_p$+S_e. The ratio of edge plane area to total surface area therefore yields

$$\frac{S_e}{S_{tot.}} = 1 - \frac{ab}{ab + (a+b)c}. \qquad (6.1)$$

The geometric area of the principal planes is 4cm², while the area of the edge planes is 0.6 cm², if samples are 1 mm thick. So the contribution of the edge planes is not larger than 13% of the overall geometrical surface area. If the samples are 100 microns thick, this contribution is only around 1.5%.
The reaction plane does not change significantly during activation. Therefore it is expected that the mass and the sample thickness decrease linearly.

Figure 6.2 displays the decrease of sample mass (in % of the sample weight before activation) versus the reaction time at temperatures between 350°C and 550°C. From the slope of the curves the burn-off rates were determined with a linear least square fit. From the burn-off rates the thickness changes were determined, which are plotted for various reaction temperatures versus the reaction time in Figure 6.3.
If the mass density of the GC is assumed to be constant throughout the sample, the thickness change can be calculated as follows:

$$\frac{dm}{dt} = \frac{\rho dV}{dt}. \qquad (6.2)$$

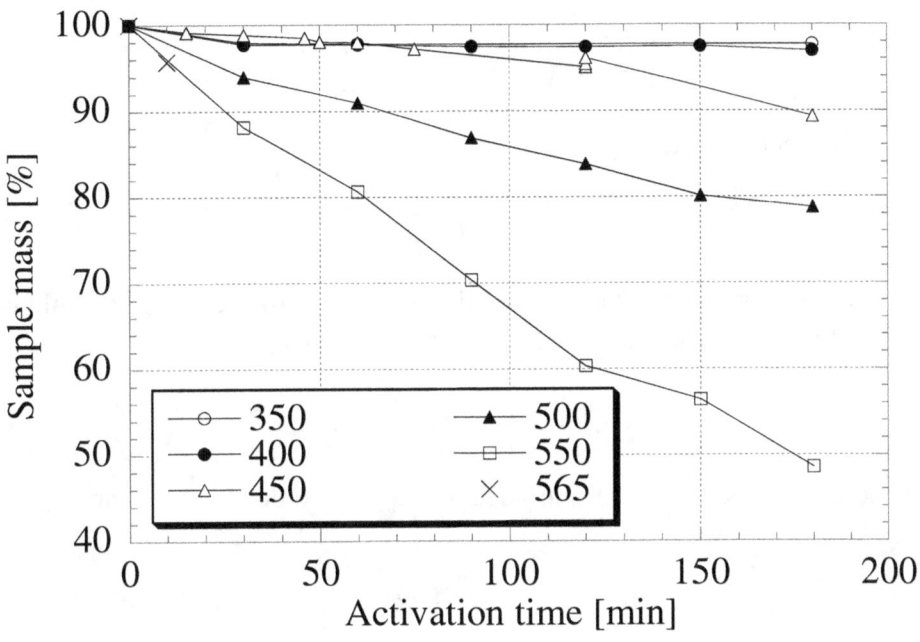

Figure 6.2: Decrease of sample mass (SIGRADUR®K with 1 mm thickness) in % during thermal activation as a function of reaction time. Numbers and symbols in the legend denote the reaction temperature in °C.

6.1. THICKNESS AND MASS CHANGES DURING ACTIVATION

The total differential of V is given by

$$dV = \frac{\partial V}{\partial a} da + \frac{\partial V}{\partial b} db + \frac{\partial V}{\partial c} dc \qquad (6.3)$$

and therefore

$$\frac{dV}{dt} = bc\frac{da}{dt} + ac\frac{db}{dt} + ab\frac{dc}{dt} \qquad (6.4)$$

The time derivatives of a, b and c are identical, because we assume that the burn-off is isotropic. One may write therefore

$$\frac{dV}{dt} = (ab + (a+b)c)\frac{dc}{dt}, \qquad (6.5)$$

which can be approximated by

$$\frac{dV}{dt} = ab\frac{dc}{dt}, \qquad (6.6)$$

because the contribution (a+b)c is a small contribution to the whole sample geometrical surface. The mass change therefore is written as

$$\frac{dm}{dt} = \frac{\rho dV}{dt} \approx \frac{\rho S_p dc}{dt}. \qquad (6.7)$$

The temperature dependence of mass decay can be well described by an exponential:

$$m(t,T) = m_0 - at \cdot e^{bT}, \qquad (6.8)$$

where m_0 is the initial mass of the sample. The time dependence of the burn-off is determined by the constant a. The sample mass and thickness exhibit a linear decrease with reaction time.

The temperature dependence of the burn-off is governed by an exponential with constant b (Arrhenius law). The burn-off therefore is a process controlled by chemical reaction. The constant b must be considered as the activation energy, which is known to be 13.8 kcal/mol for carbon [16].
 To verify that the measured data are in accordance with theory, burn-off rates a_T were converted in $mg/(cm^2 h)$, plotted in Arrhenius-representation and compared with data found in literature [26, 16].
 Figure 6.4 displays the oxidation rates of graphite and various glassy carbons as a function of the reaction temperature (Arrhenius plot).
The slope of the Arrhenius plot curves is the same for the data found in references

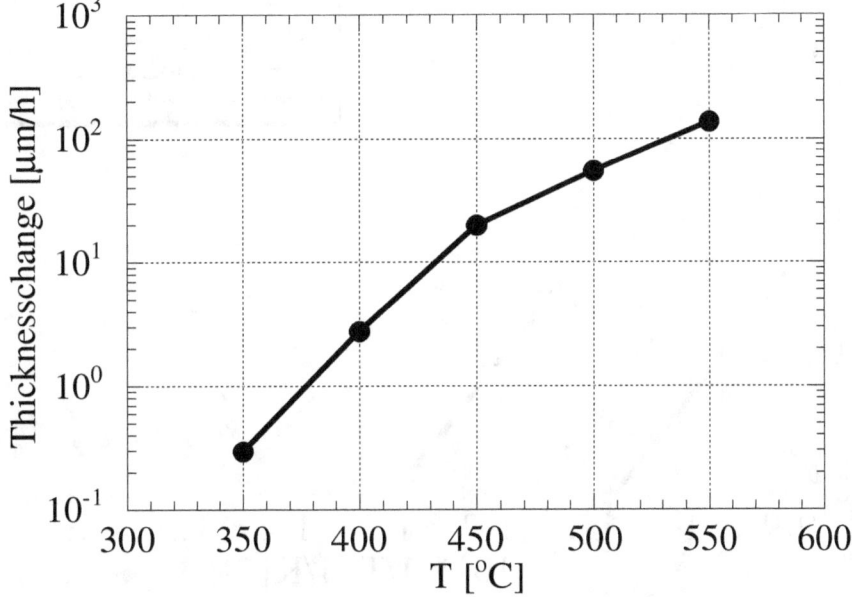

Figure 6.3: Thickness change of K 1mm GC samples as a function of reaction temperature. The data points are experimental values, obtained from the burn-off rate a_T. Note that the thickness change of the overall sample is plotted on a logarithmic scale.

6.1. THICKNESS AND MASS CHANGES DURING ACTIVATION

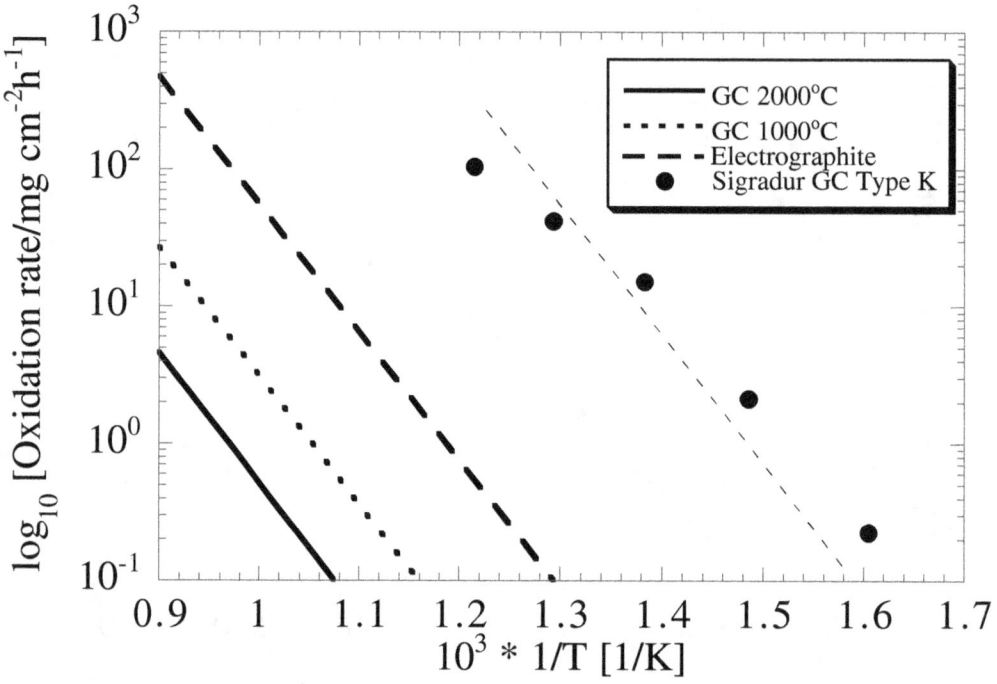

Figure 6.4: Comparison of the oxidation rates of glassy carbons and graphite. The straight lines in the left part of the plot concern data found in literature [26, 16]. Their slope corresponds to an activation energy of E_a=13.8 kcal/mol [16]. The data points on the right were obtained from the burn-off experiments on the SIGRADUR®K samples with 1 mm thickness. The dotted line serves as a guide to the eye and has the same slope as the data from literature.

[26, 16]. The samples have therefore the same activation energy of E_a=13.8 kcal/mol. The activation energy of the 1 mm SIGRADUR®K is measured to be 10.3 kcal/mol. This value does not deviate too much from the literature value. The scattering of the measured data and error bars were not taken into account. Also the data from [26, 16] have no error margins.

Adsorption of humidity

When measuring the weight of the samples on the balance it turned out that the sample weight increased during cooling down. This was probably caused by adsorption of gas and humidity from the ambient atmosphere in the active film of the samples, which leads to an increased overall weight of the samples.
The increase of weight should be directly correlated with the porosity or internal surface area of the activated samples. Samples with larger internal surface area should exhibit a more pronounced increase of weight than those samples with smaller internal surface area.
Figure 6.5. displays the weight changed after 24 hours cooling down for samples activated at 550°C as a function of reaction time.

With increasing reaction time the weight change is getting smaller, indicating that the internal surface area of these samples decreases with reaction time. It is anticipated here that this is in line with the findings from capacitance measurements at 550°C, where we observed that the DLC is decreasing correspondingly after longer oxidation times (Section 6.3.1, Figure 6.18).
It must be assumed that at 550°C activation temperature only a very thin porous film is created after initial stage activation. At 550°C the burn-off is large and after one hour of activation much of the GC material is burnt off.
Possibly structural inhomogeneities in the sample between surface and center of the sample are responsible for decreasing capacitance and internal surface area during activation.

Mass density of the active film

The density of the untreated GC is around 1.5 g/cm^3, depending on the sort of GC (pyrolysis schedule, precursor).
During gas phase reaction a film with open pores is created. The opening process of the pores is linked with a burn-off of material, which is interlinking the graphene crystallites (tetragonally coordinated carbon, or hydrogen).
The film material therefore must have a lower mass density than the original GC material with closed pores.

6.1. THICKNESS AND MASS CHANGES DURING ACTIVATION

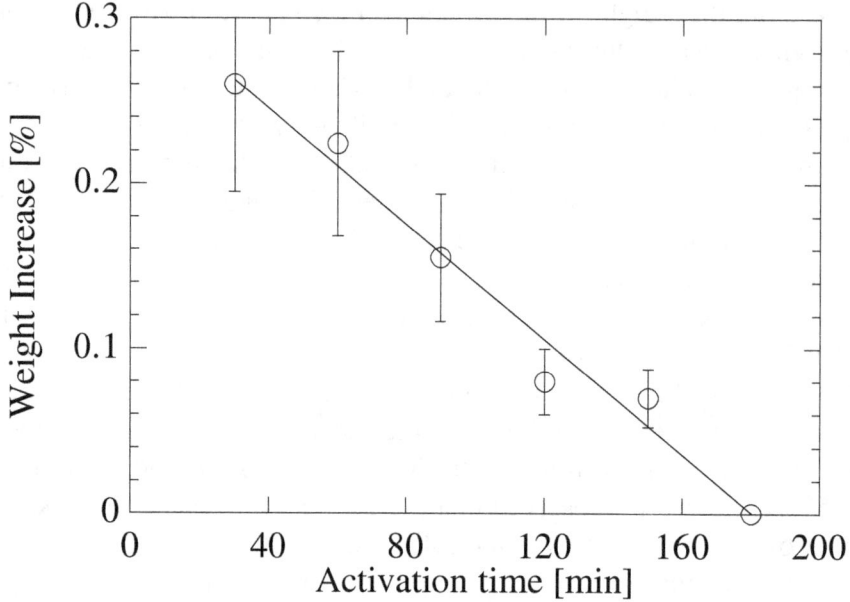

Figure 6.5: Evolution of weight increase due to humidity and gas adsorption after cooling down versus reaction time for 1 mm SIGRADUR®K, activated at 550°C. The weight of the hot sample just removed from furnace is set 100 %. The weight of the sample 24 hours later is compared with the weight mentioned before. The change is small, but significant.

As mentioned in Section 6.1.1., the mass decrease due to the thinning of the overall sample, which yields a decreasing sample thickness, can be written as a linear function of reaction time (equation 6.8).
It was particularly difficult to determine the mass density of the active film, because the film is bound to the unreacted core material, which has a mass density different from the film. Additionally, only small amounts of material were available for measurements.
Therefore indirect methods had to be applied.
The total mass of a flat monolithic GC sample with area S_p consists of the unreacted bulk mass and the active film mass:

$$m_{tot.} = \rho_{core}V_{core} + \rho_{film}V_{film} \tag{6.9}$$
$$m_{tot.} = S_p \rho_{core} d_{core} + S_p \rho_{film} 2 d_{film} \tag{6.10}$$
$$m_{tot.} = S_p((D - 2d_{film})\rho_{core} + 2d_{film}\rho_{film}) \tag{6.11}$$
$$D = d_{core} + 2d_{film} \tag{6.12}$$

V being the volume and ρ the density of each species. The overall sample thickness is D.
The mass density ρ_{core} is known by the manufacturers information and can be measured easily by its geometrical data and sample weight.
The decrease of the sample thickness is directly correlated with the mass decrease by equation 6.7 and yields a linear relation:

$$D(t) = L_0 - \alpha t, \tag{6.13}$$

L_0 being the initial thickness of the sample and α being the burn-off coefficient for the sample thinning, which can be determined directly by thickness determination of samples activated for various times.
The active film thickness d_{film} can be approximated by a square-root law, as will be shown in the next section (Model for Film Growth):

$$d_{film}(t) \approx \beta \sqrt{t}. \tag{6.14}$$

From the mass changes during activation, the active film density in principal can be derived then from following relation:

$$\frac{m(t)}{S_p} = \left(L_0 - \alpha t - 2\beta\sqrt{t}\right)\rho_{core} + 2\beta\sqrt{t}\rho_{film}. \tag{6.15}$$

Experimental data for the mass changes of SIGRADUR®K with 1 mm thickness, activated at 450°C, were fitted to above equation and plotted in Figure 6.6.
In the upper plot the film growth rate β and the film mass density ρ were indepen-

6.1. THICKNESS AND MASS CHANGES DURING ACTIVATION

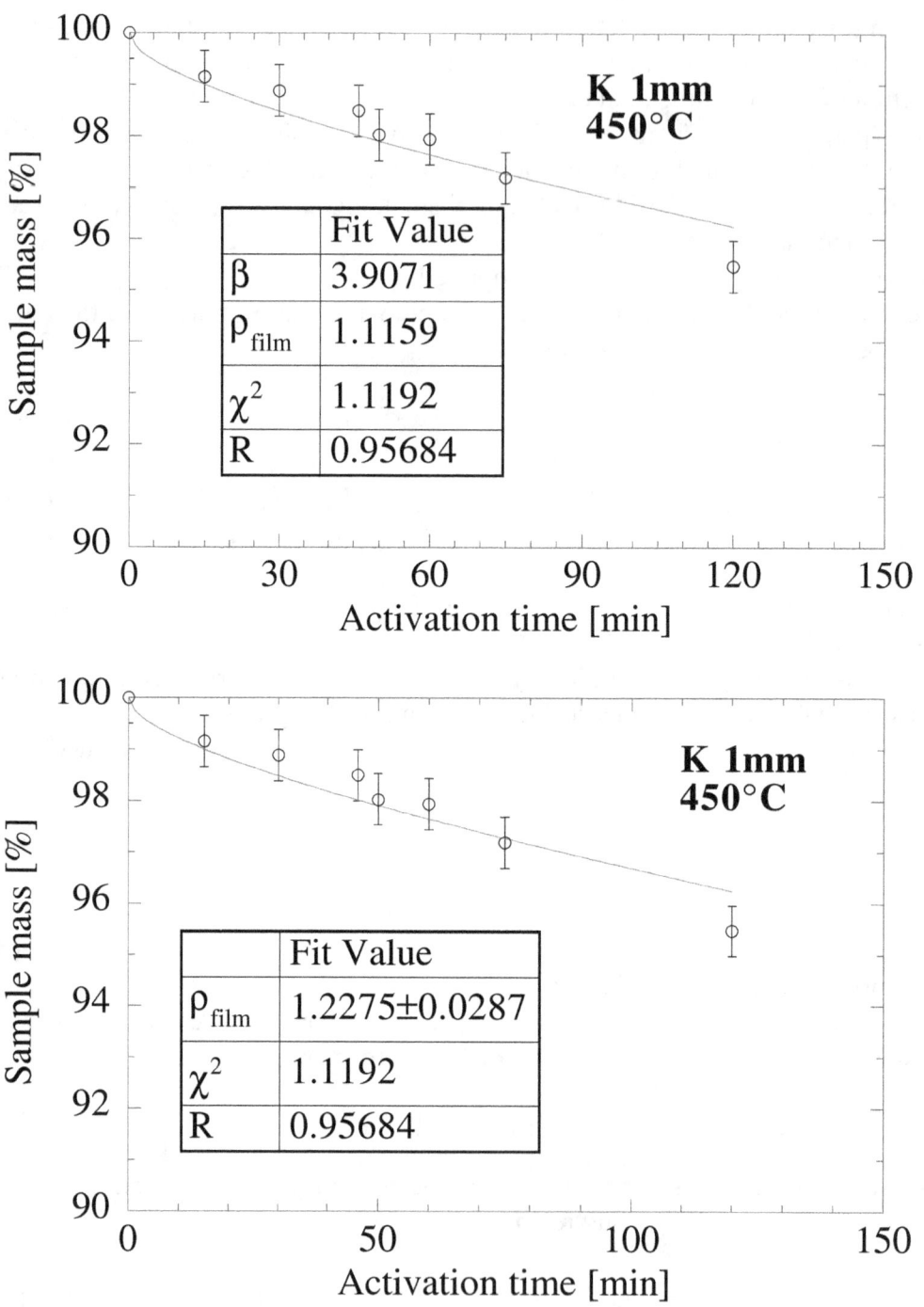

Figure 6.6: Decrease of sample mass (SIGRADUR®K with 1 mm thickness, 450 °C) during activation and fitted curve according to equation 6.15.
Top: Film growth rate β and the film mass density ρ were independently fitted.
Bottom: Only the film mass density was fitted.

dently fitted. A film mass density of 1.1 g/cm^3 and a film growth rate of 3.9 μm/h were found. In the plot on the bottom only the film mass density was fitted. A film growth rate of 5.3 μm/h was taken for fitting, as obtained by film thickness measurements (see next section). The mean square deviation of the data is χ^2, and the correlation is R.

A film mass density around 1.1 to 1.2 g/cm^3 seems rather high. But if the mass lost during film growth was larger, this should be more pronounced in the experimental data for the weight loss.

From thickness determination experiments on SIGRADUR®K samples with 60 microns nominal thickness a film mass density also around 1.1 to 1.2 was found. From film thickness determination and mass decrease measurements on SIGRADUR®K with 1 mm thickness a film mass density of 0.83 g/cm^3 was estimated (see subsection 6.3.7 *Estimation of DLC by Geometrical Considerations*.)

It was assumed that the mass density of the active film is constant. However, Raman measurements [88] give evidence that the active film has a gradient in its structure, which possibly could have its origin in a mass gradient.

It is not unreasonable to assume that the film near the reaction interface core/film has a larger mass than the film at the sample outer surface.

The profile of the Raman absorbed intensity on the sample was linear, so that one could conclude a linear mass density profile between the two interfaces mentioned. Figure 6.7 displays a schematic sketch of the distribution of the mass density versus the sample thickness line.

A linear concentration gradient usually yields a squareroot-like growth law for the growth of thin films.

In the formulation presented above (eqns. 6.9 - 6.15) the film mass can be taken into account as follows:

$$dm_{film} = \rho(z)dV(z) \qquad (6.16)$$

$$m_{film}(t) = \int dm(t) = \int_{V_{film}} \rho(z)dV(z) = S_p \int_{z_a}^{z_b} \rho(z)dz \qquad (6.17)$$

$$\text{and } z = z(t). \qquad (6.18)$$

The coordinate z denotes the thickness changes in the sample during activation, therefore being a function of activation time t: $z(t)$.

The integration has to be carried out between the core/film interface z_b and the outer surface z_a of the sample, which are functions of reaction time.

Because a sufficient number of data were not available, the film mass density investigations could not yield an exact determination of ρ_{film}. The probable value is a little above 1 g/cm^3.

It should be clarified in further experiments whether the film mass density depends

6.1. THICKNESS AND MASS CHANGES DURING ACTIVATION 93

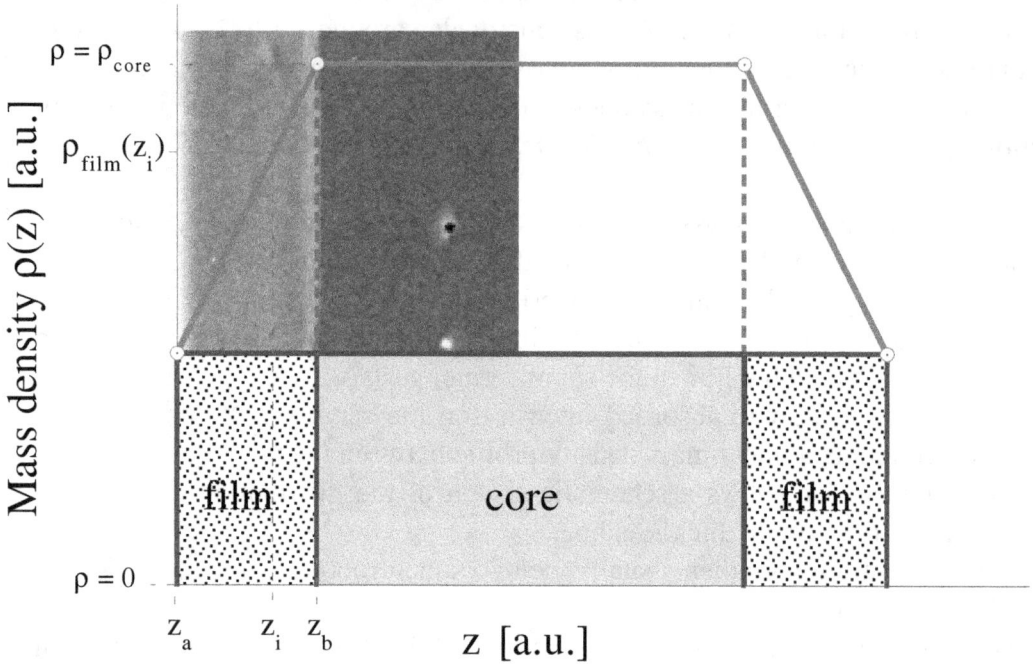

Figure 6.7: Schematic of mass density distribution in activated GC sheets. A GC sheet is displayed with unreacted core and active film on left and right side. A cut off picture of a SEM micrograph is pasted above the sketch for better view. A linear mass density gradient is assumed along the film profile. A constant mass density is assumed along the unreacted core.

on the structure of GC and also an the activation procedure. It would also be of fundamental interest whether the density is constant throughout the whole film or, what is expected by the author due to the results of the Raman experiments, the density follows a gradient which is not known yet.

Sample surface morphology

Finally it is important to know that morphological changes occur on the sample surface during activation. The GC samples as received from the manufacturer are flat and shiny and, depending on the type of GC, even of optical quality.
After some activation, the samples exhibit cracks on their surface. Upon further activation, the cracked material peels off and the surface becomes very rough. The grade of roughness depends on activation time and activation temperature.
To describe the morphological state, a *morphological phase diagram* is displayed in Figure 6.8. The diagram concerns SIGRADUR®K with 1 mm thickness. Only three levels of different morphology were used to create a grid describing the surface. The actual sample surface, however, exhibited more finesses depending on reaction time and temperature.
For low activation temperatures and short activation times the sample surface remained smooth or at least did not show cracks.
After 120 minutes of activation at 400°C cracks start to occur on the sample. The same holds for 60 minutes of activation at 450°C. Samples activated for 30 minutes at 550°C have a similar surface.
After 90 minutes of activation at 450°C material begins to peel off. Samples activated for 30 minutes at 500°C have a similar surface.
The samples activated longer than 60 minutes at 550°C have no flakes or peeled off material on the surface. However, although marked with the same symbol, their surface differed from the surface marked with the symbol for lower activation temperatures and shorter activation times.

6.1. THICKNESS AND MASS CHANGES DURING ACTIVATION

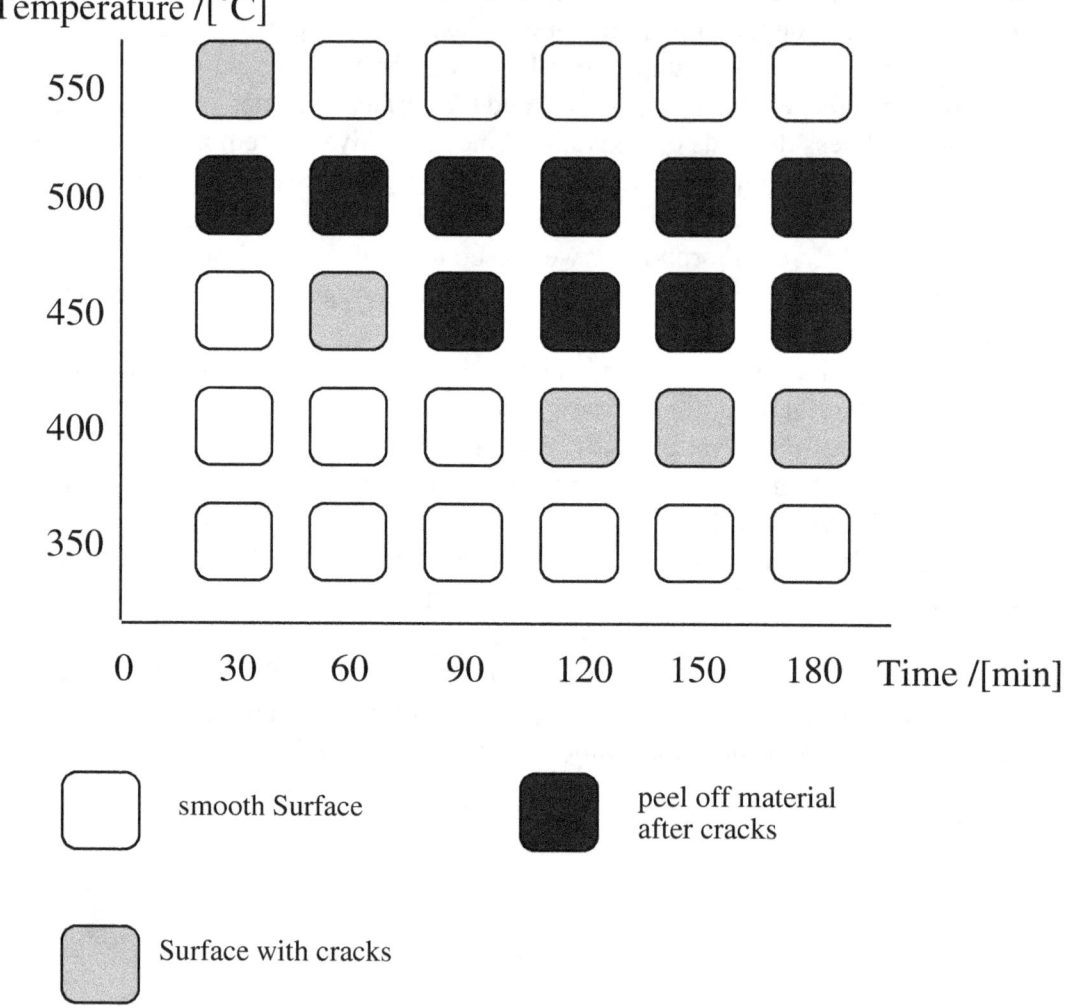

Figure 6.8: Sample surface morphology after thermal activation at different times and temperatures for SIGRADUR®K with 1 mm thickness.

6.1.2 Film Thickness

The knowledge of the thickness of the active film is of essential importance, because information such as specific capacitance or resistance are based on the active film volume only.
Only the specific data give information on the structure of the active film.
By knowledge of the film thickness, the volumetric capacitance and the volumetric internal surface area can be derived.
Experimental results found with SAXS (Pore radii distribution, internal surface area, porosity) only make sense when they can be attributed to the film material.
In the next section a model for the active film growth during thermal oxidation is developed and analytically solved. It was found that the asymptotic solution for the film thickness is a constant, i.e. the ratio of reaction rate and diffusion coefficient. The transient behaviour of the film growth is governed by a so-called [89] *Generalized LambertW-function $\mathcal{G}$*.

6.2 A Model for the Film Growth

A mathematical expression for the film thickness in a flat sample, reacting with a fluid, as a function of reaction time is derived. Major part of this section is submitted for publication in *Chemical Engineering Science* [90]. The formulation is based on a gas-solid reaction similar to the unreacted core shrinking model and takes into account two moving reaction frontiers, one due to the shrinking unreacted core, and one due to the changing thickness of the samples as a result of combustion. The equation of motion for the two reaction boundaries can be solved exactly for flat samples applying a LambertW function [91, 92, 89]. The model for the flat samples is compared with experimental results.

	NOMENCLATURE
*	concerning reaction which changes sample's overall size
$\mathcal{A}, \mathcal{A}^*$	species of fluid molecules or atoms
α	burn-off rate [μm/min]
arg(W)	argument of LambertW function
$b, (b^*)$	stoichiometric coefficient of film growth (burn-off) reaction
$\mathcal{B}, \mathcal{B}^*$	species of solid molecules or atoms
β	film growth constant $\frac{bD_{eff}C_{A_s}}{\rho_B}$ [μm^2/min]
C_1	integration constant
C_A	concentration of $\mathcal{A}$ at any site
C_{A_c}	concentration of $\mathcal{A}$ at the unreacted core
$C_{A_s}, C_{A_s}^*$	concentration of $\mathcal{A}$ at sample surface
C_{A_z}	concentration of $\mathcal{A}$ at a position z
D_{eff}	effective diffusion coefficient [cm^2/sec]
$\mathcal{G}$	generalized LambertW function
γ	abbreviation; $\gamma = \beta/\alpha$
k_s, k_s^*	rate constant of film growth reaction
$N_\mathcal{A}$	number of atoms or molecules of species $\mathcal{A}$
$N_\mathcal{B}$	number of atoms or molecules of species $\mathcal{B}$
$Q_\mathcal{A}$	flux of species $\mathcal{A}$ through area S at position z
ρ_B	mass density of unreacted bulk material [g/cm^3]
ρ_F^*	mass density of porous film [g/cm^3]
S	constant sample surface exposed to reaction [cm^2]
t, t'	reaction time [min]
Θ	film thickness [μm]
V	volume of material under investigation (bulk or film) [cm^3]
W	LambertW function
z	space coordinate in flat sample [μm]
Z_0	initial sample thickness [μm]
z_e, z_e'	sample thickness [μm]
z_i, z_i'	unreacted core/bulk thickness [μm]

6.2.1 Introduction

For the non-catalytic reaction of particles with surrounding fluid, the two most fundamental models being used are the progressive conversion model and the unreacted core shrinking model [93]. The unreacted core model assumes that the zone of reaction is narrowly confined to the interface between the unreacted solid and the product, when the chemical reaction is very rapid and the diffusion of fluids sufficiently low, for instance due to a small porosity of the unreacted material. If the solid contains enough voidage so that the fluid reactants can diffuse freely into the interior of the solid, the reactions between fluid and solid may be viewed as occurring homogeneously throughout the entire solid to produce a gradual variation in solid reactant concentration in all parts of the particle [94], which is the progressive conversion model.

Ishida, Wen and Shirai developed a more general model for solid-gas reactions, which takes into account two stages of reaction and which incorporates the progressive conversion model and the unreacted core shrinking model as limiting cases [95, 94, 96].

A shrinking particle radius was taken into account by that model. E. Comparini and R. Ricci [97] studied a general model for an isothermal noncatalytic reaction confined to the front between a fluid and a solid and proved that the solution of the problem converges to the solution of the pseudo-steady-state approximation.

Liddo and Stakgold [98] introduced a model for the combustion of a porous solid with two moving fronts, one of which is distributed through the entire porous solid and the other of which has partial conversion ahead of it. Carter [99] introduced a kinetic model for solid-state reactions and took into account that the product after reaction may have another volume than the unreacted sample. In ref. [100] the conversion-time relationship for particles with changing size and unreacted core shrinking is exercised for samples with various geometries.

Rehmat et al. [101] were the first group which took into account a shrinking particle size and a shrinking unreacted core for noncatalytic gas-solid reactions for the unsteady-state heat balance. They calculated the internal temperature of reacting samples and the conversion-time relationship numerically.

Sotirchos and Amundson [102] introduced a model for the transient combustion of porous char particles, which was able to describe the dynamic behavior of particles burning in an oxygen environment.

In technological applications usually the conversion-time relationship is of considerable interest, or models are developed in order to determine whether reactions are controlled by diffusion or reaction. However, the evolution of film thickness as a function of reaction time was merely studied.

In this work the case of a flat solid particle having two reactions with the fluid at two moving boundaries, is treated.[1] A diffusion controlled reaction creates a film in the particle according to the unreacted core shrinking model.

Another process burns off the film at the sample boundary and should be controlled by the chemical reaction. This second process leads to a shrinking of the whole particle itself. The aim of this study is to find an explicit expression for the film thickness of the flat particle as a function of reaction time and material properties and experimental conditions.

Therefore a model for combustion of particles with two moving reaction fronts is mandatory.

A model for the evolution of film thickness during reaction is presented, which takes into account a moving boundary for the unreacted core [93, 100] of the sample and a changing size of the particle size itself. It is assumed that the reaction front is confined to the particle surface and the other confined to the unreacted core surface. To the best of our knowledge, such a model was not yet established. Although the case treated in this work is restricted on a sample with decreasing thickness during reaction (e.g. by combustion), the case of a growing sample may be solved in the same way.

A real system which in fact exhibits the properties which are described in this model, is also provided. Glassy Carbon (GC) can be oxidized thermochemically with the result that a film with open pores is created on the surface of the GC [103]. The thickness of this film follows in general a growth law of the squareroot type [104, 105, 99]. The GC is also burnt off during oxidation. The burn-off and the film growth are two competing processes. The film thickness as a function of the oxidation time therefore depends on both processes.

GC samples prepared like this can be utilized as electrodes for so called supercapacitors, a novel class of energy storage devices [12, 3].

6.2.2 Establishing the Model

A flat sample with an initial thickness Z_0 is considered, as illustrated in Figure 6.9. The model may be applied to a semi-infinite plate in the halfspace $z \geq 0$ as well. At an arbitrary time t the thickness of the sample is $z_e(t) \leq Z_0$. The thickness of unreacted core material is $z_i(t)$. The film thickness is

$$\Theta(t) = z_e(t) - z_i(t). \tag{6.19}$$

First, we begin with the reaction controlled combustion of the sample occurring at its outer surface. Symbols marked with an asterisk (*) concern this reaction. The sample thickness decreases linearly as

$$z_e(t) = Z_0 - \alpha t \tag{6.20}$$

---

[1]The mathematically more difficult case for spheres is treated in Appendix B.

6.2. A MODEL FOR THE FILM GROWTH

I:
Illustration of a flat sample with initial thickness Z_0 (a) and subsequent decreasing thickness z_e during reaction for different reaction times (b, c, d).

II:
Evolution of film thickness $\Theta = Z_0 - z_i$ in a flat sample with constant thickness Z_0. The unreacted bulk material thickness z_i decreases upon reaction (b, c, d).

III:
Evolution of unreacted bulk material thickness z_i and sample thickness z_e during reactions. After short reaction time (b), there is a significant film thickness $\Theta = Z_0 - z_i$, but the shrinking of the sample is negligible. After further reaction (c), the film thickness increases and the sample shrinking is remarkable. Finally (d), reaction extends throughout the whole sample, leaving behind only film material, which is also burnt off upon further reaction.

due to burn off. For the constant α we find [93]:

$$\alpha = \frac{b^* k_s^* C_{As}^*}{\rho_F^*}, \tag{6.21}$$

where b^* denotes the stoichiometry coefficient of the burn-off combustion:

$$\mathcal{A}^*(fluid) + b^*\mathcal{B}^*(solid) \rightarrow fluid\ and\ solid\ products, \tag{6.22}$$

ρ_F^* is the mass density of the porous film, C_{As}^* the concentration of reactant $\mathcal{A}$ at the outer surface of the particle and k_s^* the rate constant of the burn-off reaction in equation (6.20).

We keep this result for α in mind for further evaluation. Note, that when for any time t the unreacted core thickness $z_i(t)$ is known, then the thickness of the film $\Theta(t)$ is also known.

Now we start with an Ansatz similar to that as exercised in [93] for particles with spherical geometry. As we will treat the problem for particles with plane geometry, equations from [93] will be changed accordingly and rewritten here for the reader. A chemical reaction is considered, which creates only the film

$$\mathcal{A}(fluid) + b\mathcal{B}(solid) \rightarrow fluid\ and\ solid\ products \tag{6.23}$$

and which is balanced by following equation:

$$-dN_\mathcal{B} = -b\,dN_\mathcal{A} = -\rho_\mathcal{B}\,dV = -\rho_\mathcal{B}\,d(z \cdot S) = -\rho_\mathcal{B}\,S\,dz, \tag{6.24}$$

N being the number of atoms or molecules of species $\mathcal{A}$ or $\mathcal{B}$, $\rho_\mathcal{B}$ being the density of the solid (unreacted core), b the stoichiometry coefficient of the film growth reaction and S being the area of the sample subjected to reaction in equation (4).

The gas $\mathcal{A}$ attacks the solid $\mathcal{B}$ and leaves behind a film and reaction products, as displayed in Figure 6.9 II. Note, that the sample overall thickness remains unchanged so far.

Use is made of the steady state assumption that the unreacted core is stationary and the reaction rate equals the flux Q_A of $\mathcal{A}$ through the reaction area S:

$$-\frac{dN_A}{dt} = S \cdot Q_A \qquad (6.25)$$

We assume that the flux Q_A depends on the concentration of fluid $\mathcal{A}$ along the film:

$$Q_A = D_{eff}\frac{dC_{Az}}{dz}, \qquad (6.26)$$

D_{eff} being the effective diffusion coefficient of reactant $\mathcal{A}$ in the film. Comparing equation (6.25) and (6.26), we may therefore write

$$-\frac{dN_A}{dt} = S \cdot Q_A = S \cdot D_{eff}\frac{dC_{Az}}{dz} \qquad (6.27)$$

and integrate along the range of film thickness:

$$-\frac{dN_A}{dt}\int_{z=Z_0}^{z=z_i} dz = S \cdot D_{eff}\int_{C_A=C_{As}}^{C_A=0} dC_A \qquad (6.28)$$

which yields

$$-\frac{dN_A}{dt}(z_i - Z_0) = -S \cdot D_{eff} \cdot C_{As}. \qquad (6.29)$$

After inserting equation (6.24) in (6.29) and some rearranging we may write

$$\int_{z_i=Z_0}^{z_i=z_i'}(z_i - Z_0)dz_i = \frac{b \cdot D_{eff} \cdot C_{As}}{\rho_B}\int_{t=0}^{t=t'} dt. \qquad (6.30)$$

Integration extends over the range of film thickness between Z_0 and z_i. The solution of this integral is

$$z_i(t) = Z_0 - \sqrt{2\beta t} \text{ with } \beta = \frac{b \cdot D_{eff} \cdot C_{As}}{\rho_B}. \qquad (6.31)$$

The film thickness therefore yields

$$\Theta(t) = Z_0 - z_i(t) = \sqrt{2\beta t}, \qquad (6.32)$$

6.2. A MODEL FOR THE FILM GROWTH

which means that for a sample of infinite thickness the film will become infinte as well, if reaction proceeds infinitely.

In equation (6.30) the assumption was made that the sample thickness remains constant during reaction, having Z_0 in the integrand instead of $z_e(t)$. This constraint is omitted now in order to allow a shrinking of the whole sample by the burn-off reaction, as experienced in our experiments with glassy carbon and illustrated in Figure 6.9 III. The burn-off reaction will be described phenomenologically with the invariant α as described in equation (6.21).

Instead a shrinking of the sample, a swelling also would be applicable. Depending on the system and experimental conditions under investigation, α then may change its sign accordingly.

We continue with equation (6.30), replacing Z_0 by $z_e(t)$:

$$\int_{z_i=Z_0}^{z_i=z'_i} (z_i - z_e) dz_i = \beta \int_{t=0}^{t=t'} dt \tag{6.33}$$

Integration extends along the film thickness from Z_0 to z_i. However, z_i is a function of time t. The integral in equation (6.33) therefore is an equation with 3 variables ($z_e(t)$, $z_i(t)$ and t). Because of

$$z_e(t) = Z_0 - \alpha t \quad \text{and} \quad dz_e = -\alpha dt \tag{6.34}$$

we may substitute the integration variable t by z_e and write

$$\int_{z_i=Z_0}^{z_i=z'_i} (z_i - z_e) dz_i = -\frac{\beta}{\alpha} \int_{z_e=Z_0}^{z'_e} dz_e. \tag{6.35}$$

Integration will not be carried out. Instead, we will continue with the following differential equation (DE), which is equivalent to equation (6.35):

$$\frac{dz_i}{dz_e} = -\frac{\beta}{\alpha} \frac{1}{z_i - z_e} \tag{6.36}$$

which is a homogeneous DE of the d'Alembert type and can be solved with the multiple valued LambertW function [106].

The LambertW function [91, 92, 89] is implicitly defined by the relation

$$W(x) \cdot exp(W(x)) = x \tag{6.37}$$

and has, for instance, following properties:

$$W(-exp(-1)) = -1 \tag{6.38}$$

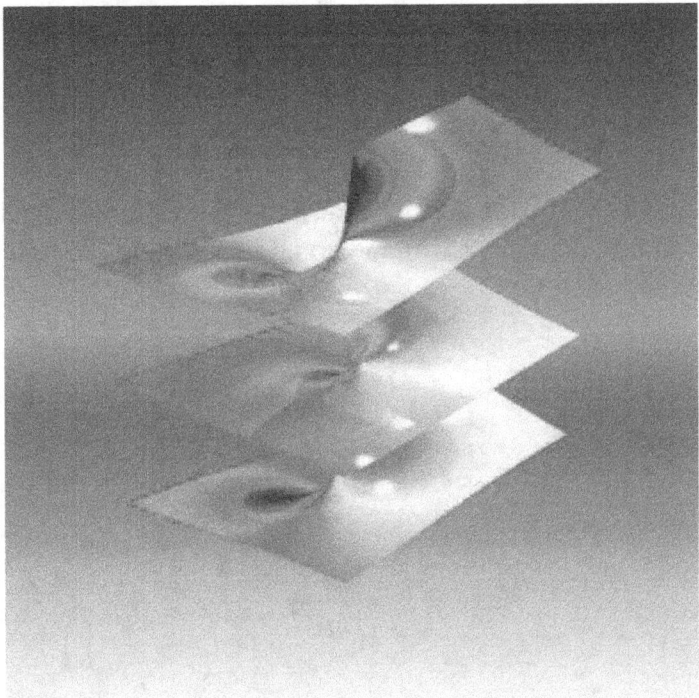

Figure 6.9: Plot of the LambertW function for various branches. LambertW is a complex valued function. Plot by courtesy of Ha Quang Le, Univ. of Waterloo, Canada.

6.2. A MODEL FOR THE FILM GROWTH

and

$$W(0) = 0. \tag{6.39}$$

Figure 6.10 displays a plot of the LambertW function with complex arguments for 3 branches.

We point out that the system treated in this work represents an exactly solvable growth problem. The solution was obtained using the Computer Algebra Program *Maple V, Release 5*.

For further investigations, only the principal branch for $x > 0$ will be considered.

$$z_i(z_e) = z_e - \frac{\beta}{\alpha}\left(1 + W\left(\frac{C_1}{\frac{\beta}{\alpha}}exp\left(-\frac{z_e - \frac{\beta}{\alpha}}{\frac{-\beta}{\alpha}}\right)\right)\right) \tag{6.40}$$

We want to find an expression for the film thickness $\Theta(t)$. Replacing z_e by its original expression (equation (6.20)) and some rearranging, the thickness of the active film is written as follows

$$\Theta(t) = z_e(t) - z_i(t) = \frac{\beta}{\alpha}\left(1 + W\left(\frac{C_1}{\frac{\beta}{\alpha}} \cdot exp\left(\frac{Z_0 - \alpha t - \frac{\beta}{\alpha}}{\frac{\beta}{\alpha}}\right)\right)\right) \tag{6.41}$$

From the initial condition $\Theta(t = 0) = 0$, we can calculate the constant C_1 in equation (6.40) and (6.41).

$$\Theta(t = 0) = 0 = \gamma\left(1 + W\left(\frac{C_1}{\gamma}exp\left(\frac{Z_0 - \gamma}{\gamma}\right)\right)\right) \tag{6.42}$$

We can rearrange this equation to

$$-1 = W(\frac{C_1}{\gamma}exp(\frac{Z_0 - \gamma}{\gamma})). \tag{6.43}$$

By comparing equation (6.38) and (6.43) we find

$$arg(W) = \frac{C_1}{\gamma}exp(\frac{Z_0 - \gamma}{\gamma}) = -e^{-1}. \tag{6.44}$$

The constant C_1 therefore yields

$$C_1 = -\gamma \cdot exp(\frac{-Z_0}{\gamma}). \tag{6.45}$$

As we now can replace the constant C_1 in equation (6.40) and (6.41), we find for the function of the active film thickness

$$\Theta(t) = \frac{\beta}{\alpha}\left(1 + W\left(-exp\left(-1 - \frac{\alpha^2}{\beta}t\right)\right)\right), \qquad (6.46)$$

or, after resubstituting for α and β:

$$\Theta(t) = \frac{b\,D_{eff}\,C_{A_s}\,\rho_F^*}{b^*\,k_s^*\,C_{A_s}^*\,\rho_B}\left(1 + W\left(-exp\left(-1 - \frac{(b^*\,k_s^*\,C_{A_s}^*)^2\,\rho_B}{b\,D_{eff}\,C_{A_s}\,\rho_F^{*2}}t\right)\right)\right) \qquad (6.47)$$

We remark that $C_{A_s} = C_{A_s}^*$, because in the case of the two combined reactions concentrations are measured at the sample surface.

Limiting cases

From equation (6.46), the following limiting cases can be derived:
(i) Considering equation (6.46) and by knowing that $W(0) = 0$ (equation (6.39)), we find that the limit of $\Theta(t)$ for large values of the oxidation time t is given by

$$\lim_{t\to\infty}\Theta(t) = \frac{\beta}{\alpha} = \frac{b\,D_{eff}\,\rho_F^*}{b^*\,k_s^*\,\rho_B}. \qquad (6.48)$$

This limit is the constant saturation value for the film thickness.
Considering the argument of the exponential in the LambertW function, we find that the saturation film thickness is reached particularly soon when the constant α is increased or the constant β is decreased. However, α dominates because it is of 2nd power in the argument of the LambertW function. Therefore, increasing the concentration of the oxidant accelerates the film growth.
(ii) If the burn-off rate α is taken zero, the expression for the film thickness with non-changing sample thickness (equation (6.31)) must be obtained.
To verify this, we use the series expansion for the generalized LambertW function [89] $G_{-1} = 1 + W[-exp(-x - 1)]$:

$$G_{-1}(x) = \sqrt{2}\,x^{1/2} - \frac{2}{3}x + \frac{1}{9\sqrt{2}}x^{3/2} + \frac{2}{135}x^2 + \ldots \qquad (6.49)$$

The expansion for the active film thickness in terms of α therefore yields

$$\Theta(t) = \frac{\beta}{\alpha}\left(\sqrt{2}\,(\frac{\alpha^2}{\beta}t)^{1/2} - \frac{2}{3}(\frac{\alpha^2}{\beta}t) + \frac{1}{9\sqrt{2}}(\frac{\alpha^2}{\beta}t)^{3/2} + \frac{2}{135}(\frac{\alpha^2}{\beta}t)^2 + \ldots\right). \qquad (6.50)$$

6.2. A MODEL FOR THE FILM GROWTH

For $\alpha=0$, only the 0th order term remains, and therefore

$$\lim_{\alpha \to 0} \Theta(t) = \sqrt{2\beta t}, \tag{6.51}$$

as expected for diffusion controlled film growth on non shrinking flat samples. It is found that the thickness of the active film followed approximately a square-root like growth law, which is expected for growth processes controlled by diffusion [105, 104, 107, 108, 99].
(iii) If the growth rate β is taken zero, no film at all is obtained:

$$\lim_{\beta \to 0} \Theta(t) = \lim_{\beta \to 0} \frac{\beta}{\alpha} + \lim_{\beta \to 0} W\left(\frac{-1}{exp\left(1 + \frac{\alpha^2 t}{\beta}\right)}\right) = 0 \tag{6.52}$$

6.2.3 Experimental

The illustration in Figure 6.9 (II) represents a realistic model of the true physical situation of the film growth, as can be seen when comparing it with micrographs of cut oxidized samples, as displayed in Figure 6.1. GC sheets of 1000 and 55 microns thickness were investigated. The thermochemical oxidation was carried out as described in Ref. [39]. For the thermochemical activation in hot air a furnace was used, the temperature of which was electronically controlled to ±5 K [39].
After a specified time, the first sample of a set was removed from the furnace, while the remaining samples of the set were kept in the furnace. Later, the second sample was removed, and so on. So a set of samples, all activated at one particular temperature (400°C, 450°C, 500°C), but for different activation times, could be provided.
Determination of film thickness was carried out with Scanning Electron Microscopy and Auger Electron Microscopy, as described in [103].
As film material pealed of from the sample due to release of elastic strain after after a specific reaction time, the theoretically expected saturation film thickness could not be measured. The thickness determination for long reaction times insofar suffers from systematic errors.
For structural information, the non-oxidized samples were investigated with X-ray diffraction (XRD) and Small Angle X-ray Scattering (SAXS). Care was taken in XRD for the position of the (002) peak, which is the only well developed x-ray peak of GC, using a silicon reference standard.

6.2.4 Results and Discussion

Figures 6.11 and 6.12 display measured data (symbols) for the film thickness and fitted curves (drawn lines) as obtained from the expression for the film thickness

Fitting Results				
Z_0 [μm]	Temperature	α_{exp} [μm/min]	β_{fit} [μm^2/min]	Θ_{fit}^{sat} [μm]
55	400°C	0.008	0.7 - 0.8	94±6
55	450°C	0.062	1.5 ±0.1	25±2
55	500°C	0.256	2 ±0.1	8±0.2
1000	450°C	0.12	19	158
1000	500°C	0.66	19	29

Table 6.1: Best fit values obtained from equation (6.47) for the film thickness of GC samples activated for different times and temperatures.

Θ (equations (6.46) and (6.47)) as a function of reaction time, burn-off rate α and film-growth rate β. Burn-off rates α were determined experimentally from the thickness of the samples after particular reaction times and particular reaction temperatures. The values for α_{exp} listed in Table 6.1 are mean values and were obtained by linear least square fits of the samples thicknesses after reaction. Only for β attempt was made for optimization. All measured data for the film thickness could be fitted using the values for α and β (Table 6.1) with reasonably accuracy. Figure 6.13 displays the film thicknesses of SIGRADUR®K (and K800 GC) with 100 microns nominal thickness (110 microns) after activation at 450°C. The thickness of the SIGRADUR®K sample is in the same range as the corresponding SIGRADUR®K with 60 microns nominal thickness. The K800 GC sample with 110 microns thickness has a larger film thickness than the corresponding SIGRADUR®K sample of comparable thickness.

No attempt was made to fit these data to equation 6.46. However, a square root function was fitted to the data from the SIGRADUR®K sample.

Temperature Dependence of Film Growth

The saturation film thicknesses Θ_{sat} of the 55 micron samples clearly decrease with increasing reaction temperature. Θ_{sat} with 500°C reaction temperature is smaller (8μm) than those obtained for the lower temperatures (25 and 100 μm).
The signature of the 2nd power of α or the reaction rate k, respectively, is also visible in the measured data and the fitted curves, because Θ_{sat} is reached earlier (at around 90 minutes) than at lower temperatures. The samples oxidized at 400°C have the largest value for Θ_{sat} (around 100 microns, although Θ_{sat} exceeds the sample thickness), but at the higher temperature (450°C) the same thickness can be reached earlier, or after same reaction time the film thickness is larger.
The same holds for the 1000 micron samples: Θ_{sat} is around 160 microns with

6.2. A MODEL FOR THE FILM GROWTH

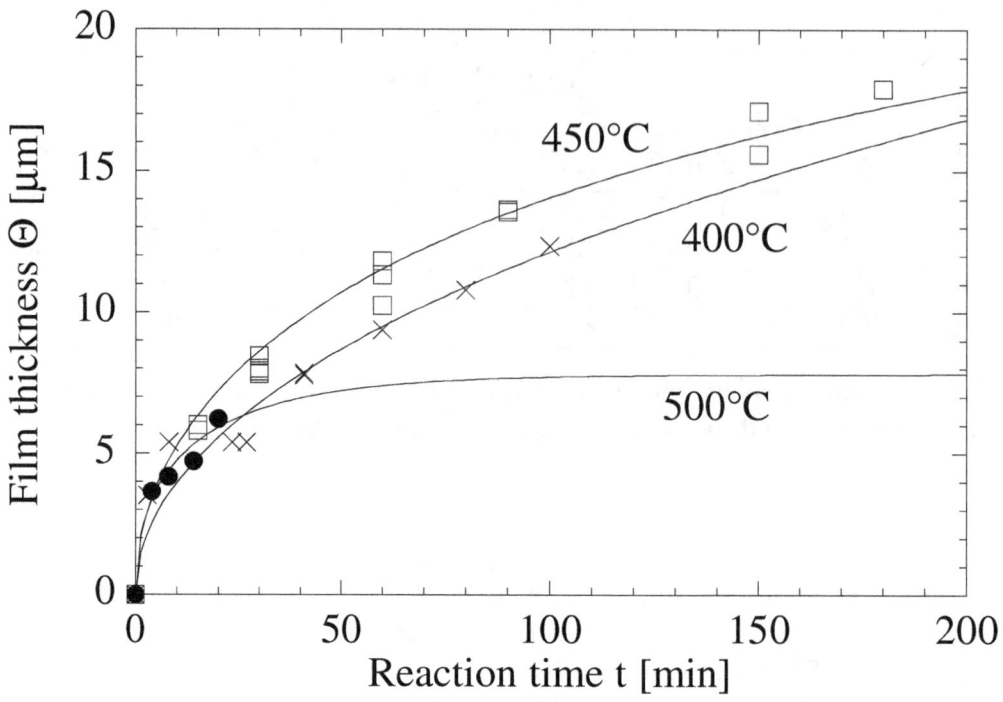

Figure 6.10: Measured (symbols) and calculated (curves) film thicknesses for the samples with 60 microns thickness after different reaction temperatures: × 400°C; □ 450°C; • 500°C.

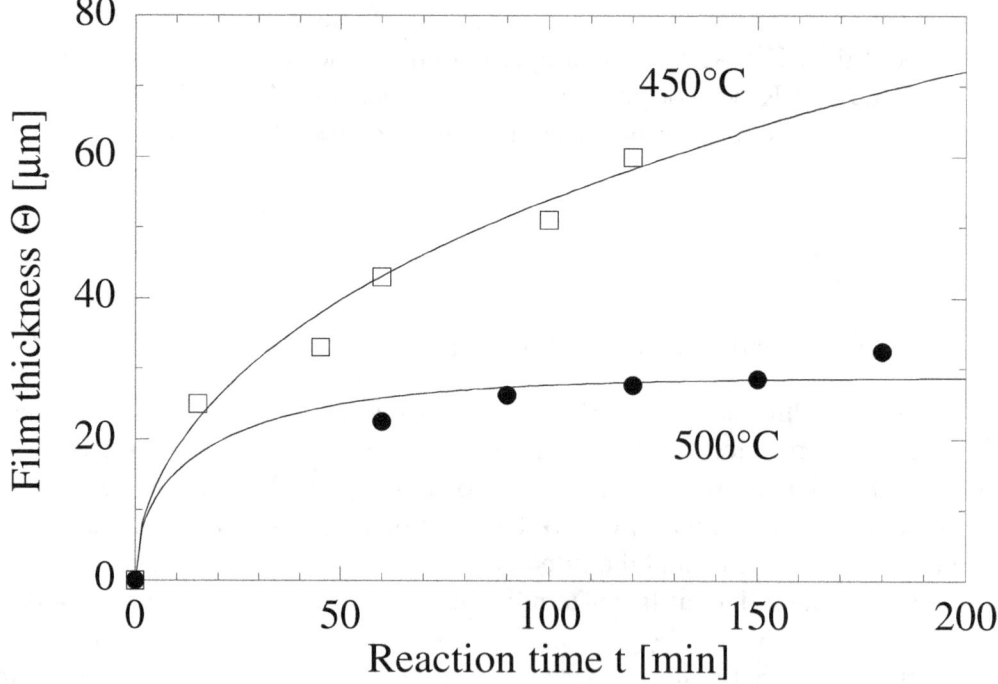

Figure 6.11: Measured film thicknesses (symbols) and fitted thicknesses (curves) for the samples with 1000 microns thickness after different reaction temperatures: □ 450°C, • 500°C.

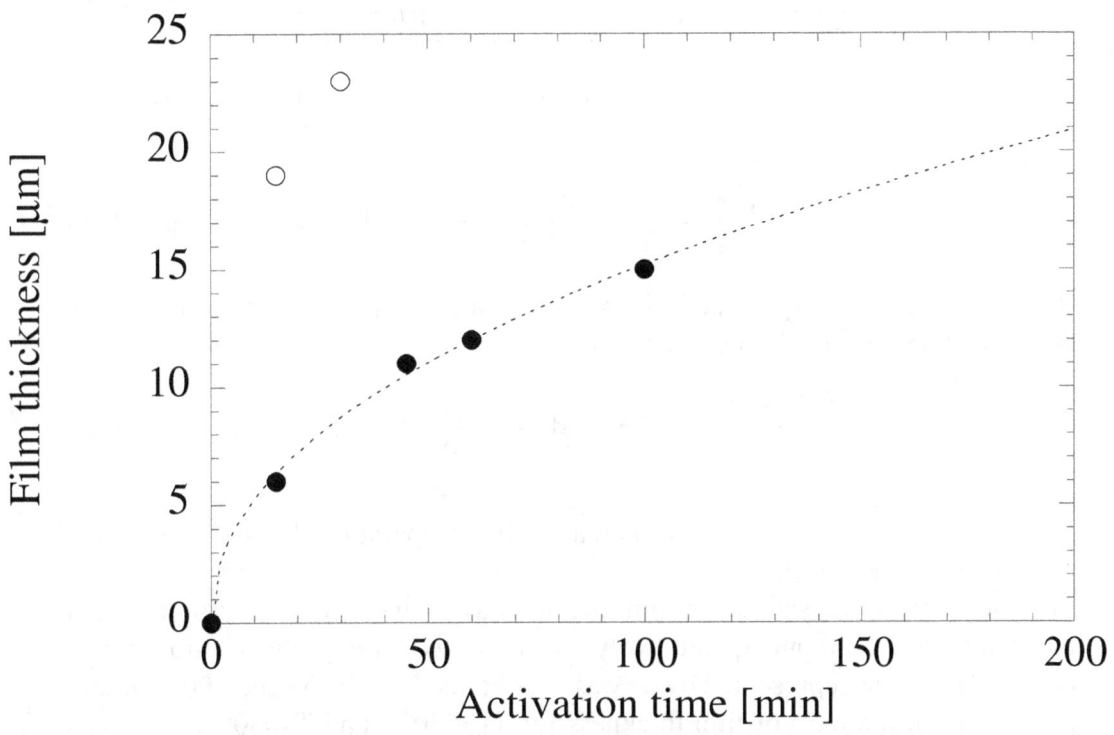

Figure 6.12: Film thickness of SIGRADUR®K (•) with 100 microns nominal thickness and K800 GC (○) with 110 microns nominal thickness after activation at 450°C. The dotted line represents a least square fit to an ordinary square-root function.

6.2. A MODEL FOR THE FILM GROWTH

Θ_{sat} reaction temperature and decreases by a factor of 5, when temperature is raised to 500°C.

As data for $\alpha(T)$ and $\Theta(t,T)$ of samples were obtained for various reaction temperatures, the temperature dependence of the film growth could be tested. It was assumed that the temperature dependence of the stoichiometric coefficients b and mass densities ρ are independent of reaction temperature within the temperature range of our experiments.

Therefore the ratio of the saturation film thicknesses of two samples activated under the same experimental conditions, but at two different temperatures T_1 and T_2, does mainly depend on the ratio of D_{eff} (represented by β) and the ratio of k (represented by α), which are both temperature dependent:

$$\frac{\Theta^{sat}(T_1)}{\Theta^{sat}(T_2)} = \frac{D_{eff}(T_1) k_s^*(T_2)}{D_{eff}(T_2) k_s^*(T_1)} = \frac{\beta(T_1)\alpha(T_2)}{\beta(T_2)\alpha(T_1)}. \quad (6.53)$$

The temperature dependence of the reaction rates k can be described by Arrhenius' law and yields

$$\frac{k^*(T_2)}{k^*(T_1)} = exp\left(\frac{E_a}{R}\left(\frac{T_2 - T_1}{T_2 \cdot T_1}\right)\right). \quad (6.54)$$

With an activation energy E_a of 54.4 kJ/mol for carbon [16] and the gas constant $R = 8.314\ Jmol^{-1}K^{-1}$, one finds

$$\frac{k^*(500°C)}{k^*(450°C)} \approx 1.8 \text{ and } \frac{k^*(450°C)}{k^*(400°C)} \approx 2.0. \quad (6.55)$$

The diffusion coefficients should increase with temperature, but only by around 10% within 50°K [100].

The evolution of the saturation film thickness as a function of reaction temperature can be discussed only qualitatively. The expected ratio for the reaction rates of around 2 (as far as represented by α) was not obtained for the 55 and 1000μm samples. Instead, the ratio of film thickness for T_1=450°C and T_2=500°C is around 5.4 for the 1000 micron samples.

For the 55 micron samples, the ratios of the reaction rates were around 8 and around 4 for the corresponding temperatures.

Using the measured values for α, the corresponding values for β were obtained by the fitting and are listed in Table 6.1.

As values for β should represent the influence of diffusion, an increase of around 10%/50°K is expected [100]. Instead, fitting yielded an increase of β by a factor of 2 for the 55 micron samples. For the 1000 micron samples, best fits were obtained for both temperatures with $\beta = 19$.

Nevertheless, there is an obvious trend that an increasing reaction rate lowers the saturation film thickness, when the diffusion coefficient is constant or less temperature dependent than the reaction rate, as stated by our model and formulation in equation 6.47. The increasing reaction rate lowers the saturation film thickness, because it is in the denominator in equation 6.47.

Deviations of the experimental results for α from Arrhenius law could maybe explained by the fact that the determination of the saturation film thickness was insufficient due to peal off of reacted material. Diffusion coefficients in porous media also cannot be determined easy [93], especially if they depend on temperature [100].
The true relations of concentrations of the reactants in the film during reaction are unknown. To quantify the influence of diffusion coefficient, especially as in porous films. Additionally it is possible that diffusion coefficients could be concentration dependent through non-stoichiometric product layers [109].
Finally we remark that structural changes occurred in our samples during reaction [103], which also could alter reaction rates and diffusion coefficients, and it was neglected that the film could have an inhomogeneous structure along the z-coordinate.

Structure Dependence of Film Growth

There is evidence from XRD and SAXS that the 55 micron and 100 micron samples have a different structure than the 1000 micron samples.
Additionally, structural changes occur in the films during oxidation [103]. From SAXS we know that the 1000 micron samples have micropores of around 12 Å diameter, whereas the 55 micron samples have a diameter of around 6 Å.
Larger pores, however, should maintain a better diffusion of reactant gases through the sample. Therefore, the film growth is facilitated for the 1000 micron samples, which can clearly be seen by comparing the saturation film thicknesses of the 1000 micron samples and the 55 micron samples.
Figure 6.14 displays XRD diffractograms of unreacted samples of 55, 100 and 1000 microns thickness. While the 55 micron sample has the maximum of the (002) peak at around 24.7°, the 1000 micron sample has the peak maximum at around 23.2°, which indicates that, after Bragg's law, the graphene sheets are closer in the thin GC samples.
The full width at half maximum of the XRD diffractograms of the thin samples is also somewhat smaller than that of the 1000 micron samples, which indicates, after the Scherrer formula, that the graphene sheets of the thin samples are laterally more extended than those of the 1000 micron samples.
We remark that the mass density of the 55 micron samples was 1.63 g/cm^3, while

the 1000 micron sample was 1.53 g/cm^3, which indicates that the 1000 micron samples have a higher porosity than the thin samples.

So a possible explanation for the differences in film growth could be that the thin samples have a smaller porosity, and therefore the transport of oxygen to the reaction front is smaller than in the case of the 1000 micron samples, with the result that the saturation film thickness is also smaller.

The correlation of structure and film thickness becomes also clear by comparison of the film thicknesses of the samples shown in Fig. 6.13. The sample pyrolyzed at 800°C has different structure than the sample pyrolyzed at 1000°C (Figures 6.33 and 6.34 in Section 6.34).

6.2.5 Conclusions

A model for the diffusion controlled growth of a film on flat samples with changing size during reaction was developed and mathematically formulated. The model represents a growth problem and was solved analytically and exactly. The film thickness depends mainly on diffusion coefficient, reaction rate and reaction time and can be expressed by an analytical equation.

The oxidation of Glassy Carbon is introduced as a process which can be described with the model, although the model is not restricted to this system nor restricted to systems with decreasing particle size. The experimental data could be fitted with reasonable accuracy.

Differences in the saturation film thickness of different samples could be qualitatively explained with their structural differences. Differences in the saturation film thickness of samples with same structure, but with different oxidation temperatures could be qualitatively explained by changes in the reaction rate and diffusion coefficients.

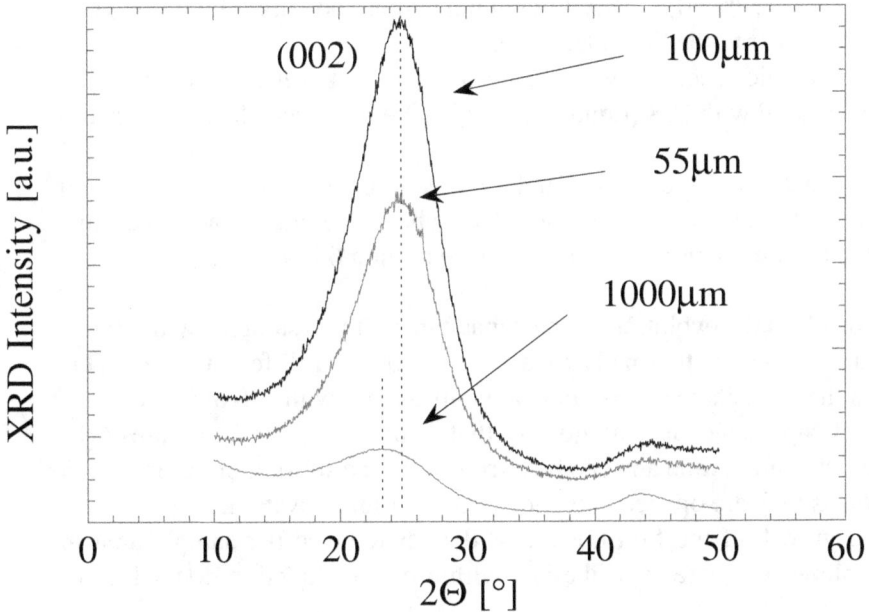

Figure 6.13: X-ray diffractograms of SIGRADUR®K with 1000 μm (upper curve) and 55 μm thickness (lower curve). For clarity, a dashed curve denotes the maximum of the (002) peak of the 55 μm sample.

6.3 Influence of Activation Parameters on Electrode Performance

6.3.1 Influence of Reaction Time and Temperature

Capacitance

Cyclic voltammograms (CV, see Figure 5.7 and Figure 6.45) of the activated GC samples exhibited a broad current wave from redox couples of the quinone/hydroquinone type around 0.4 V [78, 79, 80, 81, 82, 83], but not at other potentials between 0.0 V and 1.0 V with respect to the SCE electrode.
As no further faradaic reactions were detected by CV, it was assumed that the capacitance measured with EIS (Equation 5.42) at 0.9 V reflects the actual capacitance.
The capacitance measured at 0.4 V is usually larger than the capacitance measured at any other potential between 0.0 V and 1.0 V. The exceeding capacitance must be attributed to the Redox processes taking place around 0.4 V.

Figure 6.15 displays the evolution of the capacitance of GC samples with 60 microns nominal thickness [2], thermally oxidized at 450°C for different times. This capacitance is the capacitance per sample area, measured with EIS at 0.9 V.
The increase of capacitance is analogous to that of the film growth (Figures 6.11 and 6.12) with activation time and can be attributed to the wetting of pores, which occurs as soon as they are opened and wide enough after activation.
Similar to the growth of the film thickness, the capacitance per sample area in Figure 6.15 follows a square-root like law with a best-fit function $d(t) \approx 14.25 \cdot \sqrt{t}$.
The activation times larger than 180 minutes represent a stage of activation, the capacitance of which is extended throughout the whole sample. This was justified by a test when a drop of water was dripped on the sample top, the sample bottom immediately became wet.
Therefore, after 180 minutes of activation no non-activated GC material is present and separating the films anymore. Evidence for this was also found by SEM.

As the film thickness for these samples was known, the capacitance per apparent sample area was related to the film thickness. Thus, the volumetric capacitance was obtained, as displayed in Figure 6.16.

---

[2]Despite the nominal thickness of 60 microns according to the manufacturers information, the actual thickness of the samples prior to activation was 55 microns and accordingly less for activated samples. When the actual thickness is of no importance for interpretation of data, only the nominal thickness will be mentioned.

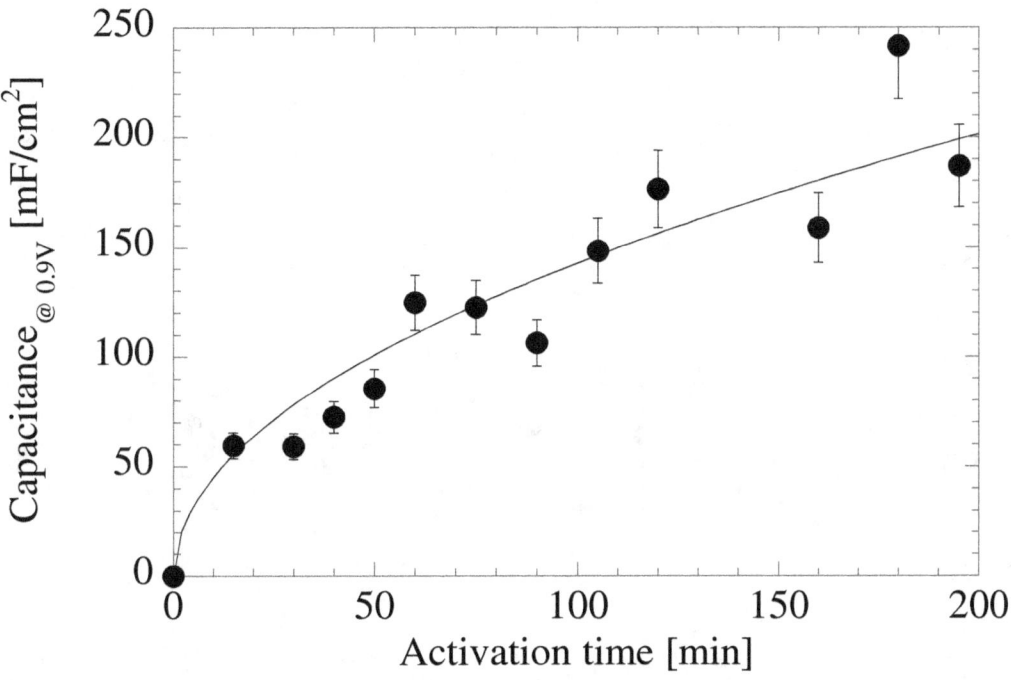

Figure 6.14: Capacitance of activated SIGRADUR®K with 60 microns nominal thickness, measured at 0.9 V bias voltage for various activation times. The activation temperature was 450°C. The values measured at samples activated for 180 and 195 minutes are divided by 2, because no active material separated the film material on each sample face. The solid line represents a least square fit of the data for a square-root function.

6.3. INFLUENCE OF ACTIVATION PARAMETERS ON ELECTRODE PERFORMANCE

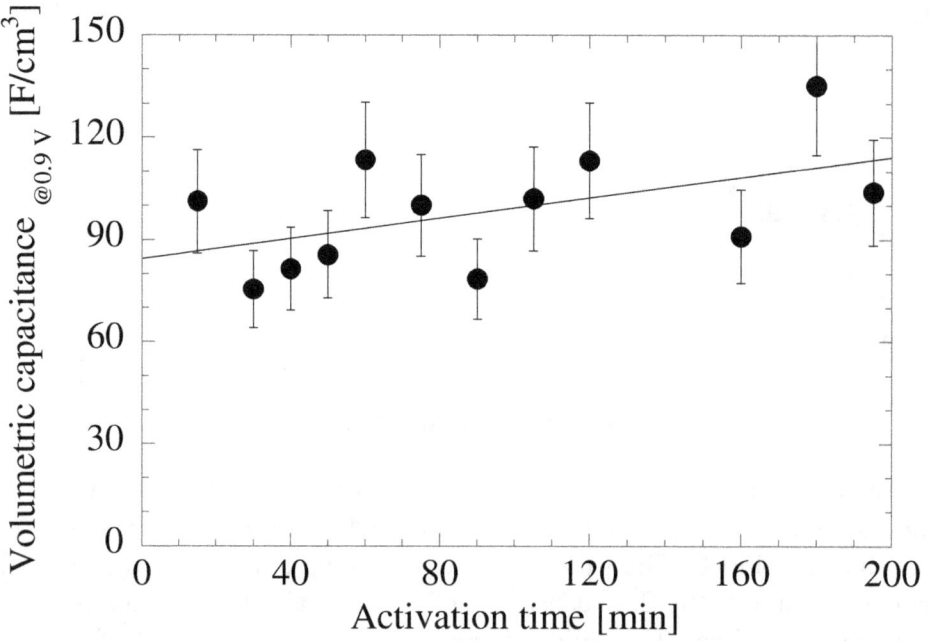

Figure 6.15: Volumetric capacitance of activated SIGRADUR®K with 60 microns nominal thickness vs. activation time, measured at 0.9 V bias voltage. The fitted curve exhibits a slight linear increase of the values of around 10% per hour on an average.

Although zigzag-shaped, the evolution of the volumetric capacitance of the samples with 60 microns exhibits an average linear increase of around 9 F/cm^2 per hour.[3]

The zig-zag shape is a common observation which occurs on monolithic samples with cracking and peeling off of activated film material. This may result in a lack of capacitance per sample area or film volume.

Figure 6.17 displays the capacitance per sample area of GC samples with 1 mm thickness, pyrolyzed at 1000°C (SIGRADUR®K) and thermally oxidized at 450°C for different times.

The capacitance increases steeply during oxidation with a rate of around 10 mF/min

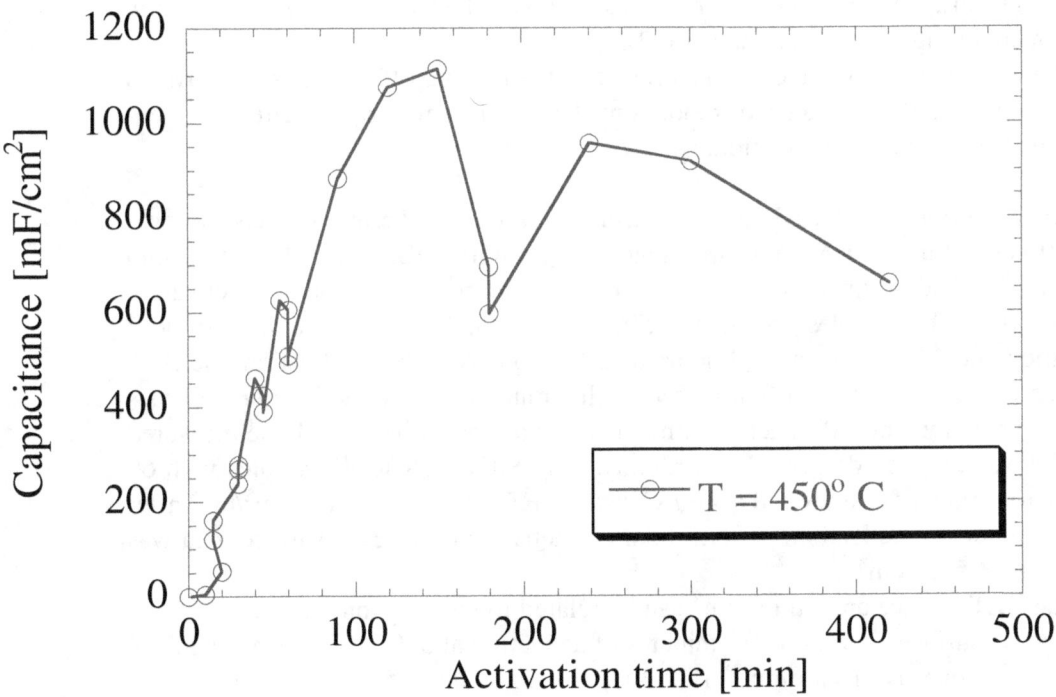

Figure 6.16: Capacitance of thermally oxidized SIGRADUR®K with 1 mm thickness at 450°C for different oxidation times. The data were obtained using EIS at a bias voltage of 0.4 V.

within the first 150 minutes.

The steep increase of the capacitance with activation time is a result of the steep increasing film thickness during oxidation. As the active film grows with oxidation, the accessible internal surface area increases accordingly.

---

[3]The increasing volumetric capacitance is a result of the opening or widening of pores during activation.

6.3. INFLUENCE OF ACTIVATION PARAMETERS ON ELECTRODE PERFORMANCE

After two hours of oxidation, the capacitance does not increase anymore, and the maximum of the capacitance of around 1100 mF/cm^2 is reached. By comparing Figures 6.15 and 6.17 for the saturation values for the capacitance and Figures 6.11 and 6.12 for saturation film thicknesses, the accordance between film growth and increasing capacitance becomes particularly clear.
The increasing capacitance is observed for various types of GC (K-type of 800°C, 1000°C and 1250°C, G-type of 2000°C) at various activation temperatures.
In terms of diffusion limited film growth it is expected that the capacitance reaches a saturation level similar to the film thickness, as considered in Chapter 6.2.
After 150 minutes of activation, an absolute maximum of capacitance (Figure 6.17) is observed, which is not in accordance with the theory of monotonous film growth during activation (Chapter 6.2).
The zigzag-shape of the curve is a result of the pealing off of flakes of activated material at different stages of oxidation. This is the same zigzag effect as mentioned before in this subsection.

The evolution of the BET internal surface area during thermal reflects the evolution of film thickness and capacitance [65]. SIGRADUR®K disks with 1 mm thickness had an internal surface area of 3.3 m^2/cm^3 after 3 hours of activation at 400°C. The samples activated at 500°C had 3 m^2/cm^3 after one hour of activation and 3.8 m^2/cm^3 after 3 hours of activation. At 550°C activation, the BET area was only around 2 m^2/cm^3 after one hour and 3 hours of activation. At times when the cracking-off of active film material was very pronounced, the measured BET surface area decreased accordingly. The SIGRADUR®K samples with 60 microns thickness, activated at 400°C, had a BET surface area of 1.3 m^2/cm^3 after 30 minutes of activation. After 3 hours of activation, the BET surface area was around 1.5 m^2/cm^3.
The BET surface area in m^2/cm^3 can be related to the film thickness for the volumetric surface area. After 30 minutes of activation at 500°C (K 1mm), the BET surface area was 1160m^2/cm^3 (1h: 1300; 2h: 1050; 3h: 1150).

For longer oxidation times the capacitance decreases.
One possible explanation for the decrease of capacitance at larger activation times could be that deeper regions of the GC are encountered, which have a structure different from the regions near the sample surface.
Regions with a different structure may have different film growth rates and also different pore sizes. Possibly the film growth rate is smaller in deeper regions of the GC. Therefore the film thickness at a particular time is smaller and also the capacitance.
In general it should be expected that monolithic GC has a gradient in its structure and therefore also a gradient in all other properties which are structure dependent.

Such inhomogeneous behaviour was reported concerning GC samples with 5 mm thickness [110].

The capacitance depends also an the oxidation temperature.
Figure 6.18 displays capacitance data of K 1mm GC, activated in air at different temperatures.
The highest capacitance is obtained at 450°C, at least within 180 minutes of ac-

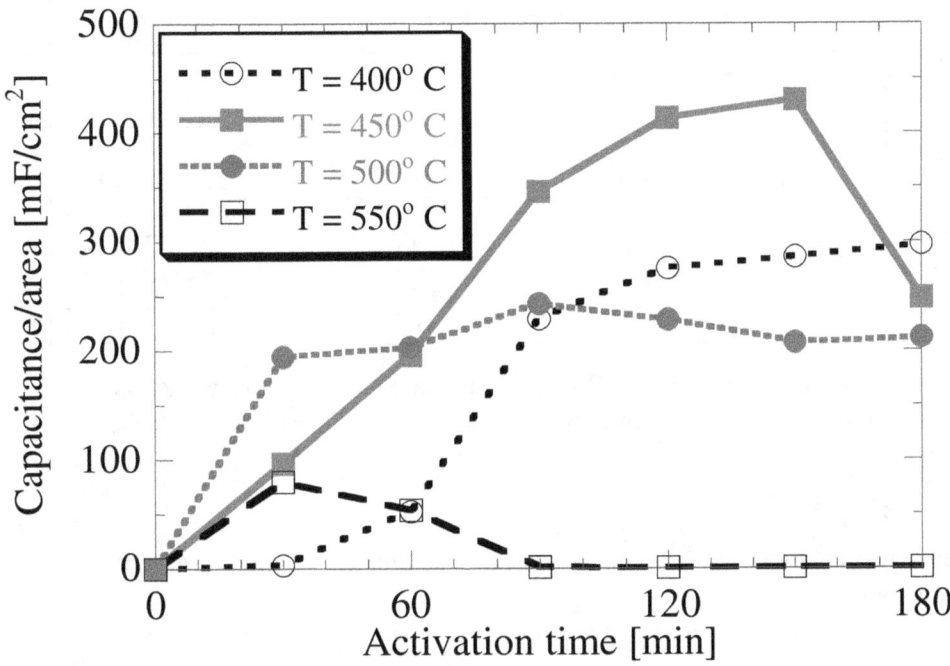

Figure 6.17: Capacitance of SIGRADUR®K with 1 mm for various activation temperatures. Data were obtained with EIS at 0.9 V bias voltage. Activation temperatures are denoted accordingly in the legend.

tivation. At 500°C the capacitance after 30 minutes activation is higher than the capacitance of 450°C after the same activation time.
These observations are in line with the active film thickness determination and with the film growth model presented in this thesis (see Chapter 6.2).
It remains open whether for activation times longer than 180 minutes at 400°C a higher capacitance is obtained. According to the film growth model, however, it is expected that the saturation film thickness after activation at 400°C is larger than at 450°C. Therefore, also a larger capacitance would be expected.
Table 6.2 summarizes results found for the capacitance at 0.4 V bias voltage at various oxidation temperatures and times for the SIGRADUR®K with 1 mm thick-

6.3. INFLUENCE OF ACTIVATION PARAMETERS ON ELECTRODE PERFORMANCE

ness.

An activation at 350°C does not yet yield a large capacitance after short activa-

	Capacitance [mF/cm^2]							
Time [min.]	350°C	400°C	450°C	460°C	470°C	480°C	500°C	550°C
10	-	-	3.63	-	-	-	-	-
20	-	-	53.42	-	-	-	-	-
30	-	12.4	238	-	-	-	195	161
40	-	-	461	-	-	-	-	-
55	-	-	625	-	-	-	-	-
60	-	222	606	-	-	-	520	119
90	1.8	421	883	602	-	-	600	3.44
120	2.9	514	1076	-	844	635	539	1.04
150	9	610	1114	-	-	-	552	1.89
180	76	298	695	-	-	-	522	1.58

Table 6.2: Capacitance of SIGRADUR®K, 1 mm thickness for various activation times and temperatures. The samples were measured with EIS at 0.4 V bias voltage and 0.1 Hz.

tion, because it takes more than two hours to obtain a value of around 10 mF/cm^2 for the capacitance.

Increasing the activation temperature accelerates the evolution of capacitance essentially, which is mainly caused by the evolution of film growth.

After the same activation time at 450°C the capacitance is increased by two orders of magnitude, the effect of which is qualitatively in accordance with the findings for the film growth. The maximum capacitance is reached between 120 and 180 minutes of activation at 450°C, as can be seen in Table 6.2. A capacitance of around 1200 mF/cm^2 is expected for 0.4 V bias potential. As capacitance measured at 0.4 V and 0.9 V differs by a factor of two, the maximum capacitance at 0.9 V should be around 600 mF/cm^2. The film thickness should be in the range of 60 microns after activation at 450°C for 120 to 180 minutes. The volumetric capacitance therefore yields 100 F/cm^3. The energy density of the corresponding active film with 1 Volt potential yields 50 kJ/liter [4]. It is recalled that the 5 Volt capacitor stack prototype (Figure 1.1) has an energy density of 0.6 kJ/liter.

During activation at 500°C the saturation value for the capacitance of around

---

[4]A capacitor made from this film material also has current collectors, separators, electrolyte etc. (Figure 2.4), which has to be taken into account.

550 mF/cm^2 is reached after one hour and maintains its value for at least up to three hours after beginning of activation.

With 550°C activation temperature the maximum value of the capacitance (around 160 mF/cm^2) is already reached after 30 minutes. After 90 minutes, the capacitance is only around 1 mF/cm^2. Note that the weight increase of the samples activated at 550°C decreases with activation time (Section 6.1.1., Figure 6.5), which is in line with the decreasing capacitance.

The decreasing value for the maximum capacitance with increasing activation temperature supports the theoretical predictions that the reaction rate suppresses the film growth and therefore the evolution of capacitance.

The shift of the maximum on the transient towards shorter activation times supports the assumption that structural inhomogeneities exist in the GC, depending on the depth reached.

The critical depth is reached sooner (due to burn-off), when the activation temperature is raised.

Thin SIGRADUR®K sheets of 60 and 100 micron thickness and disks of 1 mm thickness were thermally activated at various temperatures. According to the manufacturers specification, the sheets and disks were prepared from the same polymer and were pyrolyzed at the same HTT as the K 1mm GC samples.

The evolution of the capacitance is analogous to the evolution of the film thickness during oxidation for all different types of GC. The capacitance per film volume of the 60μm samples during the first 200 minutes of activation is around 80 F/cm^3 to 120 F/cm^3 with an increase of around 0.15 F/cm^{3}min^{-1}. The 100μm samples have an increase of the volumetric capacitance from around 90 to 130 F/cm^3 with an increase of around 0.26 F/cm^{3}min^{-1} during the first 150 minutes of activation. The samples with 1 mm thickness have a volumetric capacitance also from 90 to 130 F/cm^3, but with an increase of around 0.37 F/cm^{3}min^{-1} during the first 60 minutes of activation.

Therefore the capacitance also depends on the type of GC under consideration.

Differences in the volumetric capacitance are small, but obvious. These differences must have their origin in the different structure of the GC.

Diffusive Resistance of the Active Layer

The diffusive resistance R_{Diff} [86, 85] of the activated GC samples was determined using EIS, as described in Chapter 5.3.3, Figure 5.12.

Figure 6.19 displays impedance spectra in Nyquist representation, obtained from SIGRADUR®K samples with 60 μm thickness and activation at 450°C.

For the determination of R_{Diff} only the impedance curve itself, but not the position of the curve on the real axis of the impedance plot, is important.

6.3. INFLUENCE OF ACTIVATION PARAMETERS ON ELECTRODE PERFORMANCE

As the active film thickness increases during activation, R_{Diff} related with the sample area should also increase as well.
Figure 6.20 displays R_{Diff} of SIGRADUR®K with 60 microns thickness, which was activated at 450°C.
For short activation only, R_{Diff} is very high.
After one hour of activation, a minimum value is obtained, and with further activation the resistance increases steadily, which must be assigned to the growth of the active film.
The lowest value for the diffusive resistance of the samples with 60 μm thickness, activated at 450°C, was found to be 190 mΩcm^2 after one hour of activation.

Figure 6.21 displays the data for the diffusive resistance of SIGRADUR®K samples of 100 μm thickness, activated at 450°C for various activation times.
The trend for decreasing resistance during initial activation is observed. The minimum resistance is between 100 and 200 mΩcm^2. An increasing resistance with activation time is not yet observed within the first 140 minutes of activation, probably because scattering of the data overshadows this effect yet.

The specific diffusive resistance can be obtained by division with the film thickness. The minimum value obtained for the SIGRADUR®K with 60 microns thickness is around 200 Ωcm.
As can be seen from Figure 6.21, subsequent reduction of the samples does not alter the resistance significantly. Change of the bias potential also has no influence on the resistance.
While the capacitance depends on the bias potential (as a factor between two and three may be between the capacitance measured at 0.4 and 0.9 V), the diffusive resistance does not change significantly with the bias potential, but remains constant for all potentials.
The time constant $\tau = RC$ can be calculated from the data in Tables 6.2 and 6.3. For instance, after 55 minutes of activation at 450°C the film of activated SIGRADUR®K has τ=12 msec. After 150 minutes of activation at the same temperature the time constant increases by a factor of six to 72 msec [5]. The energy density was found to be 50 kJ/liter for this film. By division by the time constant, the power density of the active film yields around 0.7 MW/liter. The 5 Volt capacitor stack (Figure 1.1) has a power density of 38 kW/liter.

The thin samples of the SIGRADUR®K have a higher diffusion resistance (between 100 and 200 mΩcm^2) than the 1 mm SIGRADUR®K, activated at 450°C

---

[5] Mainly due to the increasing film thickness, which, however, does not increase by a factor of six.

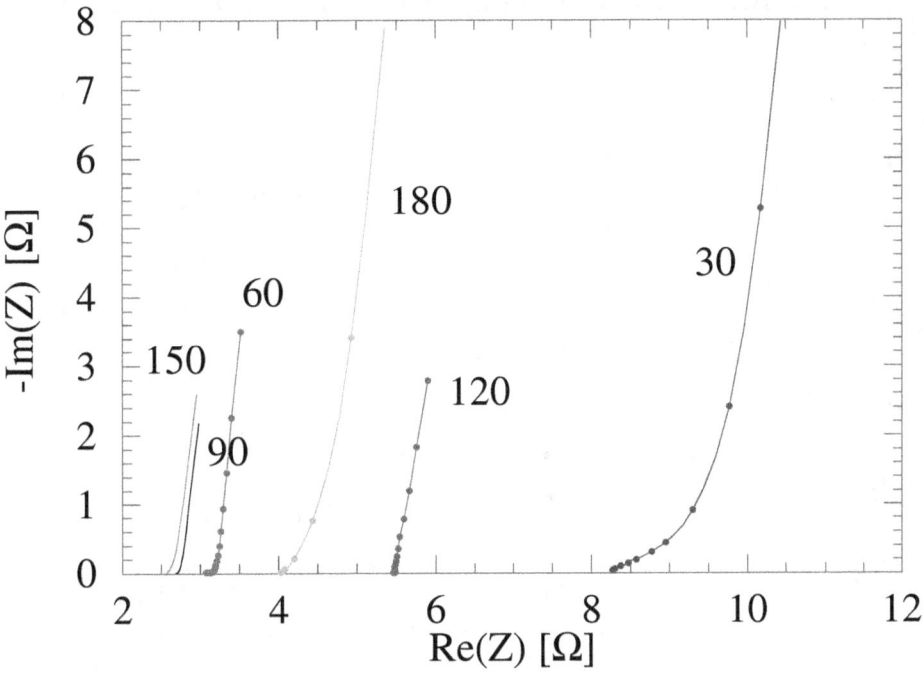

Figure 6.18: Impedance spectra (Nyquist-Plot) of SIGRADUR®K with 60μm, activated at 450°C and measured at 0.9 V bias potential. The activation time is denoted in minutes accordingly.

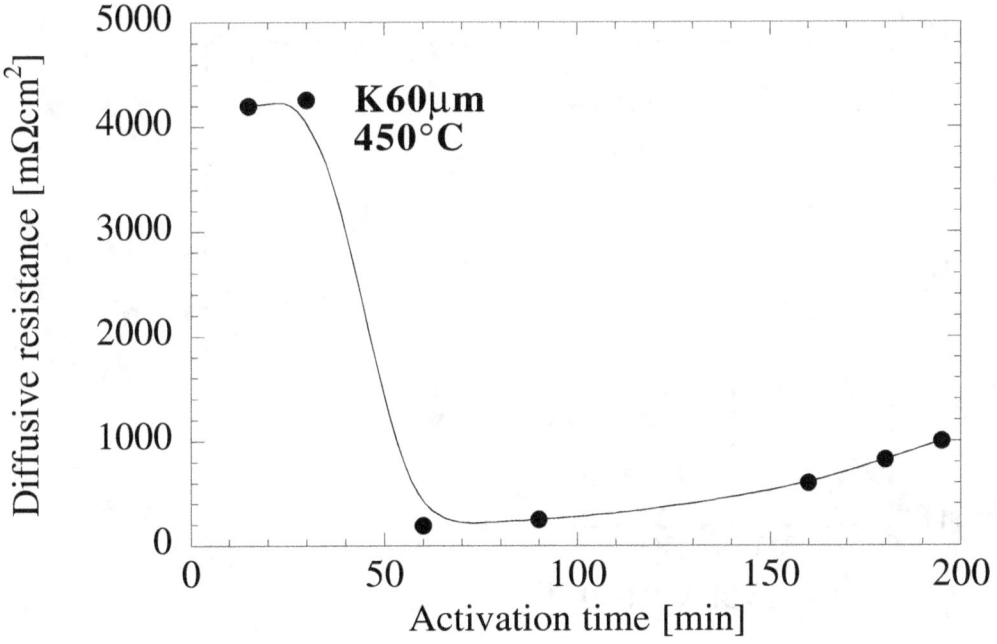

Figure 6.19: Evolution of the diffusive resistance R_{Diff} for SIGRADUR®K with 60 μm thickness, activated at 450°C. The drawn line is just a smooth interpolation of the data and serves as a guide to the eyes. The samples were measured with a bias potential of 0.9 V.

6.3. INFLUENCE OF ACTIVATION PARAMETERS ON ELECTRODE PERFORMANCE

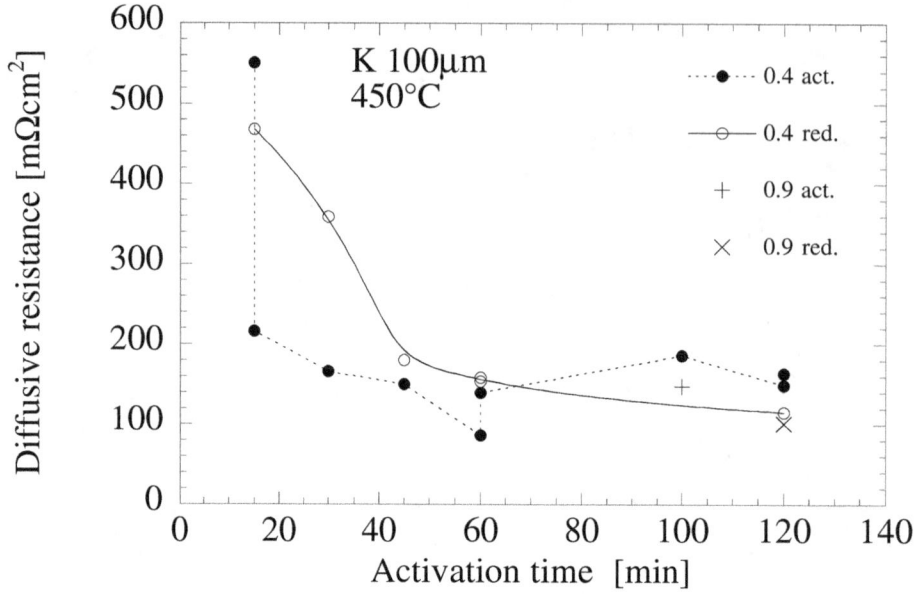

Figure 6.20: Evolution of the diffusive resistance for SIGRADUR®K with 100 μm thickness, activated at 450°C. Data points are shown for activated and also for the subsequently reduced samples, for bias potentials of 0.4 V and 0.9 V.

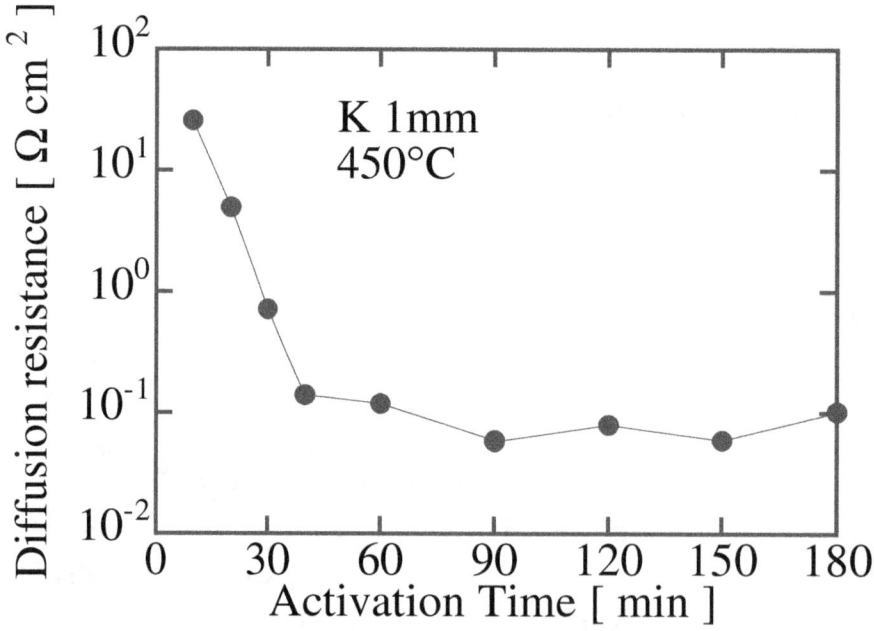

Figure 6.21: Evolution of the diffusive resistance for SIGRADUR®K with 1 mm thickness, activated at 450°C. Data points are shown for activated and also for the subsequently reduced samples, for bias potentials of 0.4 V and 0.9 V.

(between 40 and 120 mΩcm^2), as can be derived from Figure 6.22, which displays the diffusive resistance of SIGRADUR®K samples with 1 mm thickness.
After 3 hours of activation, no increase of capacitance with the film thickness is yet observed.
Unlike the samples with 1 mm and 100 μm thickness, the trend of increasing resistance with increasing activation time (= increasing thickness) was obeyed clearly only by the samples with 60 microns thickness.

Figure 6.23 displays the diffusive resistance of activated GC from pyrolyzed Capton foil, which was activated at 450°C.
The increase of the resistance with increasing activation time is expected, because the active film increases, and the diffusive resistance is a function of film thickness.
Therefore it may be assumed that the film is in so far homogeneous [6]. Electrochemically activated SIGRADUR®G has a quite lower constant ratio R/C $\approx$ 0.05 - 0.1.
 As no samples of this type of GC were activated shorter times than 30 minute, the high values for R_{Diff} were not measured yet.
Nevertheless it is an immanent behaviour of the thermally activated GC samples, that they have a very high diffusive resistance after short activation time.
Table 6.3 displays the diffusive resistance of 1 mm SIGRADUR®K with respect to the sample area (in mΩcm^2), thermally activated at different temperatures and times.
 It is obvious that the diffusive resistance is higher for samples, which were activated for short times only.
While, for instance, a sample activated 10 minutes at 450°C has a resistance of 22 Ωcm^2, samples activated around one hour have a resistance around 100 mΩcm^2 or even less.
This trend also holds for samples activated somewhat longer times at lower temperatures.
Samples activated at higher temperatures exhibit a very low resistance even after short activation.
Probably the *extent of activation* plays here an important role for the resistance at initial stage.

Apart from the diffusive resistance, more information on resistance can be derived from the impedance spectra, when a series of samples is measured.
In Figure 6.19 impedance spectra of SIGRADUR®K with 60 microns nominal

---

[6]Evidence for inhomogeneities in film properties, in particular in film structure, were however found by SAXS measurements and RAMAN investigations.

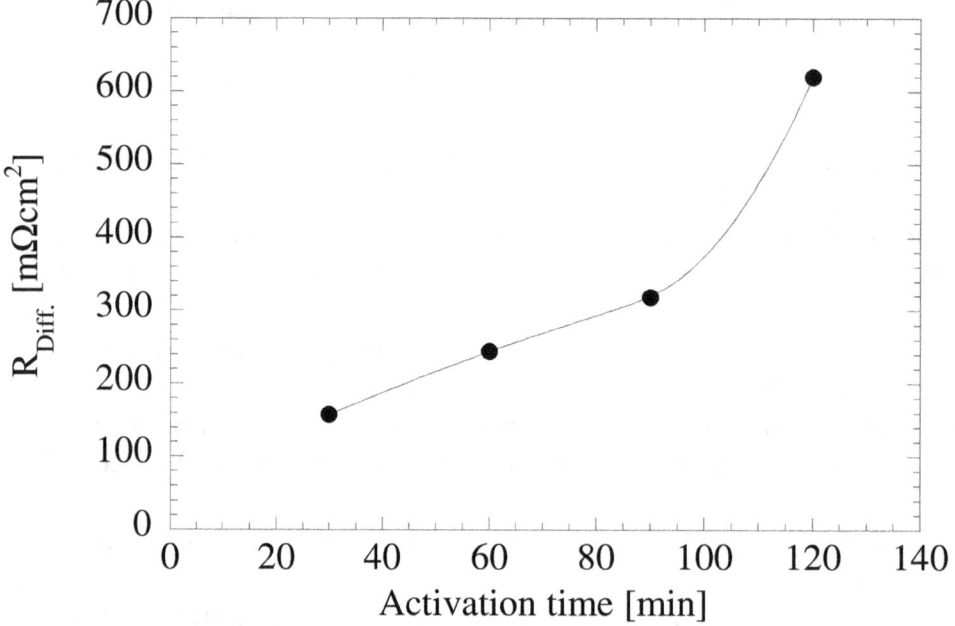

Figure 6.22: Diffusive resistance of thermally activated (450°C) SIGRADUR®K with 150μm thickness, obtained from pyrolysis of Capton sheets. The inset displays the capacitance of the samples, measured at 0.4 V bias potential. The ratio R/C is around 1 for all activation times.

	Diffusion resistance [mΩcm^2]							
Time [min.]	350°C	400°C	450°C	460°C	470°C	480°C	500°C	550°C
10	-	-	22000	-	-	-	-	-
20	-	-	5015	-	-	-	-	-
30	-	-	722	-	-	-	192	40
40	-	-	140	-	-	-	-	-
55	-	-	42	-	-	-	-	-
60	-	2680	120	-	-	-	102	33
90	20000	696	60	70	-	-	91	3.5
120	43630	309	80	-	65	82	92	7.7
150	4740	150	60	-	-	-	78	5.4
180	3150	118	140	-	-	-	181	5.6

Table 6.3: Diffusion resistance of SIGRADUR®K, 1 mm thickness, measured with EIS at 0.4 V bias voltage.

thickness were displayed. For instance, the sample activated only 30 minutes, is located at around 8 Ω on the real axis of the impedance. The intercept on the real axis therefore has a value of 8 Ω, which has to be assigned to cables, electrolyte, and GC matrix resistance.
It was observed that the samples activated only short time have a higher such resistance than the samples activated longer times. Possibly this could be interpreted as an *electric contact problem*, because the pores are at initial stage of activation too small to be wetted, and therefore no electrochemical double layer is built up. Also the behaviour of a porous electrode is not so much pronounced.

The intercept of the samples activated longer times increased with activation time. This behaviour was particularly pronounced for the sheet samples of 60 and 100 microns thickness, but not investigated systematically. Probably the thinning of the samples yielded a higher materials resistance, because less well conducting matrix material is present with proceeding activation.

Comparing the resistances and capacitances (Tables 6.2 and 6.3) we find that the capacitance of the samples activated at 400°C and 450°C increases, while the resistance decreases.
Although such an effect is throughout desirable, it cannot be understood so easily. An increasing capacitance should be caused by either an increasing film thickness or by an increasing internal surface area per film volume.

6.3. INFLUENCE OF ACTIVATION PARAMETERS ON ELECTRODE PERFORMANCE

The diffusive resistance must also increase with increasing film thickness, when the materials resistance of the non-activated GC is smaller than that of the activated GC.

One possible explanation for the decrease of diffusive resistance with activation time could be that the pore structure is altered during activation, maybe because small pores coalesce to larger pores due to activation.

This would contradict the assumption, that the porous film is homogeneous.

Frequency Response of the Capacitance

Due to the diffusion limitation at increasing frequencies, the capacitance is a function of the ac potential frequency.

All capacitance data mentioned before were obtained at the lowest applied frequency of 0.1 Hz. For higher frequencies, the capacitance has smaller values, because the transport resistances increase.

Figure 6.24 displays the capacitance of activated SIGRADUR®K samples as a function of the ac frequency.

The data clearly indicate that after three hours of activation the highest capaci-

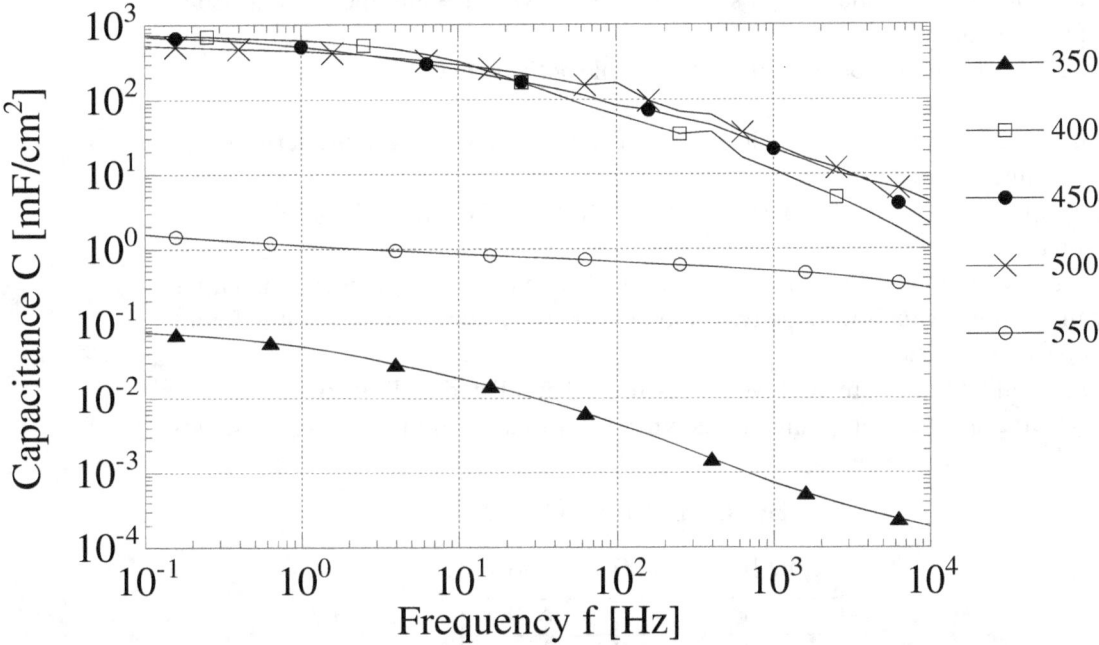

Figure 6.23: Capacitance of SIGRADUR®K with 1 mm thickness, activated 180 minutes at various activation temperatures as a function of the ac frequency. Data were obtained with EIS at 0.4 V bias voltage. Temperatures are denoted accordingly with symbols in the legend. Note that the capacitance and frequency are plotted on logarithmic scales.

tance is obtained at temperatures between 400°C and 500°C. The maximum capacitance is obtained at 450°C after 3 hours.

A temperature of 350°C is too low to create a film with sufficient thickness within three hours of activation.

An activation temperature of 550°C burns off the film too soon, so that no considerable film thickness is obtained.

6.3. INFLUENCE OF ACTIVATION PARAMETERS ON ELECTRODE PERFORMANCE

When the frequency is increased from 0.1 Hz to 100 Hz, the capacitance of the samples decreases by one order of magnitude, with the exception of the sample activated at 550°C. This is a remarkably feature of active films obtained from oxidation at high temperatures.

The feature that the capacitance remains as high as possible for increasing frequencies will be termed qualitatively *good frequency response* furtheron.

The intersection of the extrapolated tangents from the low frequency and the high frequency intercepts is related to the time constant $\tau = RC$.

As thin films should have a lower diffusive resistance, also the sample activated 3 hours at 350°C should have a small diffusive resistance and therefore a better frequency response.

However, this rule is not obeyed by the sample activated at 350°C.

Figures 6.25 display the capacitance vs. the frequency of samples activated at temperatures from 400°C to 450°C.

The capacitance of the samples activated at 400°C is larger, the larger their activation time was.

The frequency range the capacitance of which is most *stable*, however, is larger for the samples which were activated longer times. Their bending point is found at higher frequencies.

The bending points were qualitatively estimated and listed in Table 6.4.

Regarding the bending points, it becomes clear that the frequency response de-

Bending point [Hz]		
Time [min.]	400°C	450°C
30	-	≈ 0.5
60	≤ 0.1	18
90	0.7	20
120	1 - 2	30
150	4	30
180	4	30

Table 6.4: Bending point of SIGRADUR®K, activated at 400°C and 450°C. The position of the bending point is a measure for the frequency stability of the capacitance. The longer the GC activated, the more stable is the capacitance.

pends on activation temperature and activation time. This behaviour is fully in line with the results found for the capacitance and the diffusive resistance.

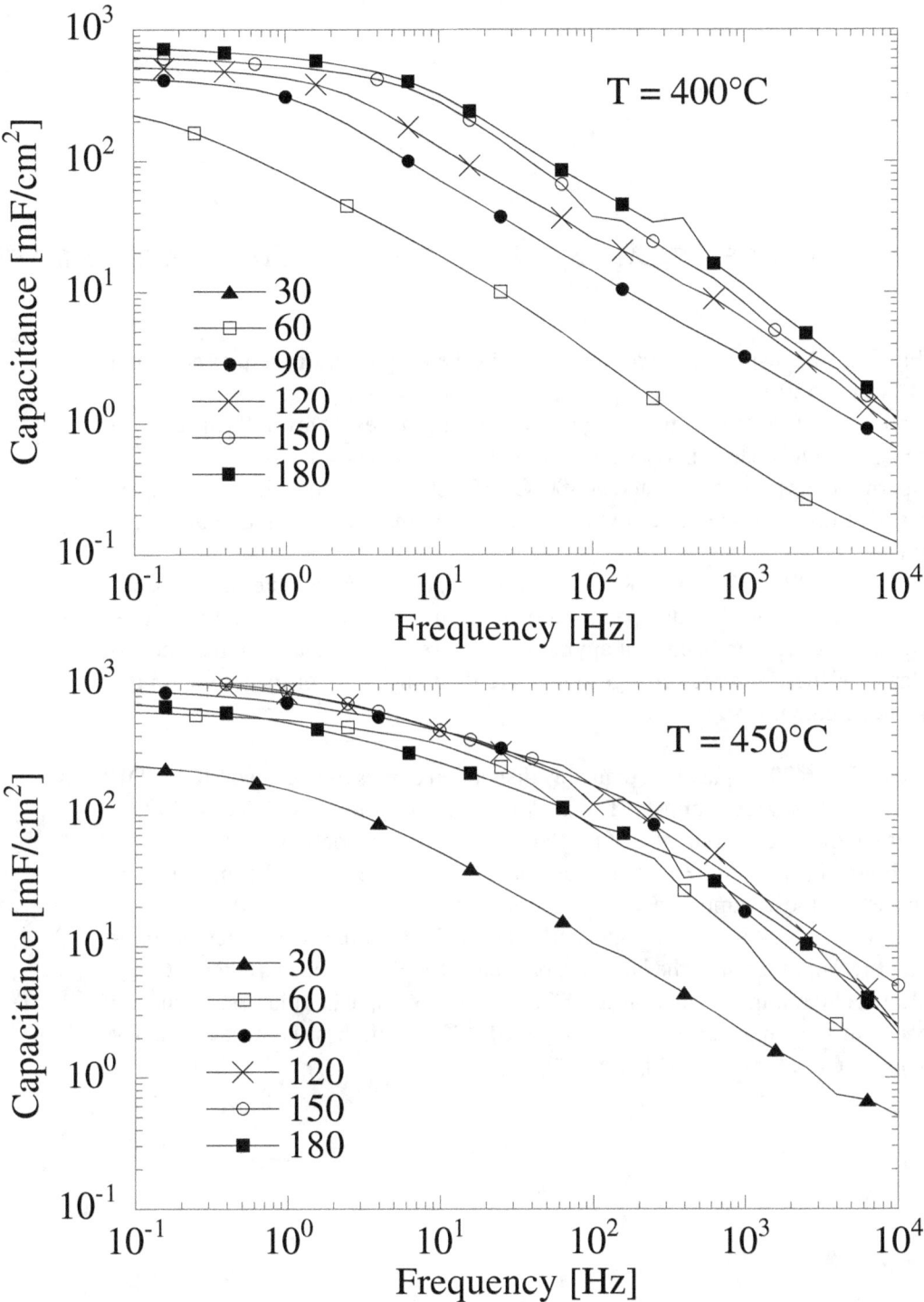

Figure 6.24: Frequency response of GC samples activated at 400 (top) and 450 °C (bottom).

6.3. INFLUENCE OF ACTIVATION PARAMETERS ON ELECTRODE PERFORMANCE 133

Figures 6.26 display the capacitance versus the frequency of samples activated at 500°C and 550°C.
The burn-off of the film at these high temperatures makes the overall capacitance decrease, which is particularly visible at the low frequencies, i.e. at 0.1 Hz.
Comparing the samples activated at 400°C and 500°C, one finds that the samples activated at lower temperatures exhibits are sharper and sooner bending of the capacitance.
An activation of the SIGRADUR®K at 550°C longer than 60 minutes is unreasonable, because no significant capacitance is measured anymore, probably because no valuable active film is present anymore. However, it is remarkable that the capacitance of these samples changes not more than one order of magnitude in the frequency range investigated.

Figures 6.27 - 6.29 display the frequency dispersion curves of activated SIGRADUR®K samples and K800 GC samples. The activation temperature was 400°C to 500°C.
The frequency behaviour of K800 after 150 minutes oxidation at 400°C is remarkably bad (Figure 6.27). For frequencies larger than 20 Hz, the capacitance is lower than of those samples activated shorter times.
After oxidation at 450°C, the regular SIGRADUR®K samples activated 90 minutes and 150 minutes have the largest capacitance for all frequencies (Figure 6.28). While after 90 minutes oxidation at 500°C the K800 sample has the best frequency response of the capacitance, the SIGRADUR®K sample has the best frequency response after 150 minutes (Figure 6.29).

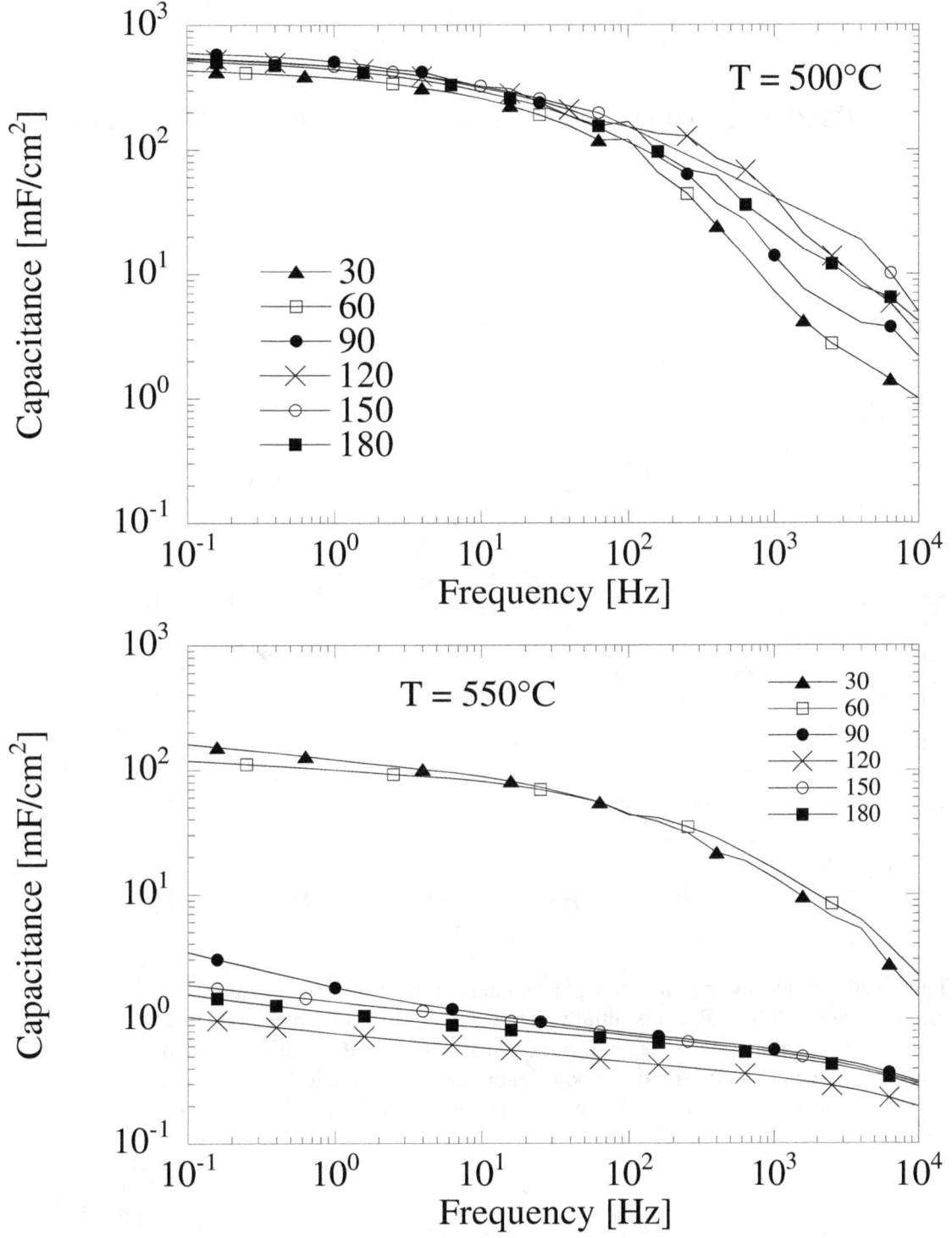

Figure 6.25: Frequency response of GC samples activated at 500 (top) and 550 °C (bottom).

6.3. INFLUENCE OF ACTIVATION PARAMETERS ON ELECTRODE PERFORMANCE 135

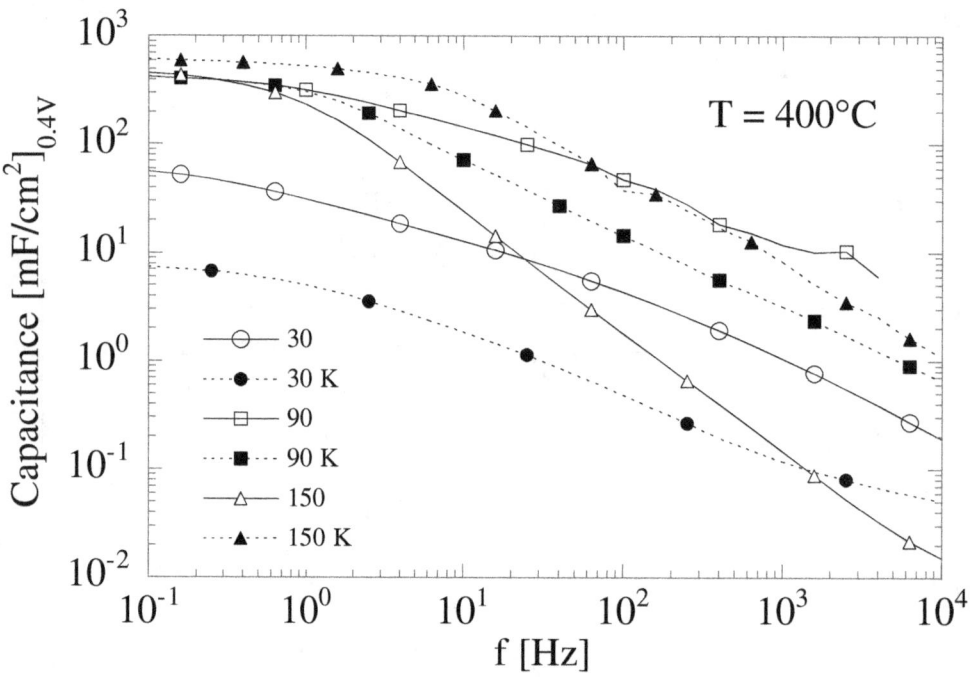

Figure 6.26: Frequency response of the capacitance of K800 GC with 1 mm thickness, oxidized at 400°C. Bias potential was 0.4 Volt. Activation times in minutes are denoted accordingly in the legend. For comparison, the SIGRADUR®K samples are displayed as well. Filled symbols denote the SIGRADUR®K, open symbols denote the K800 samples (30 K means: 30 minutes oxidized, SIGRADUR®K samples.).

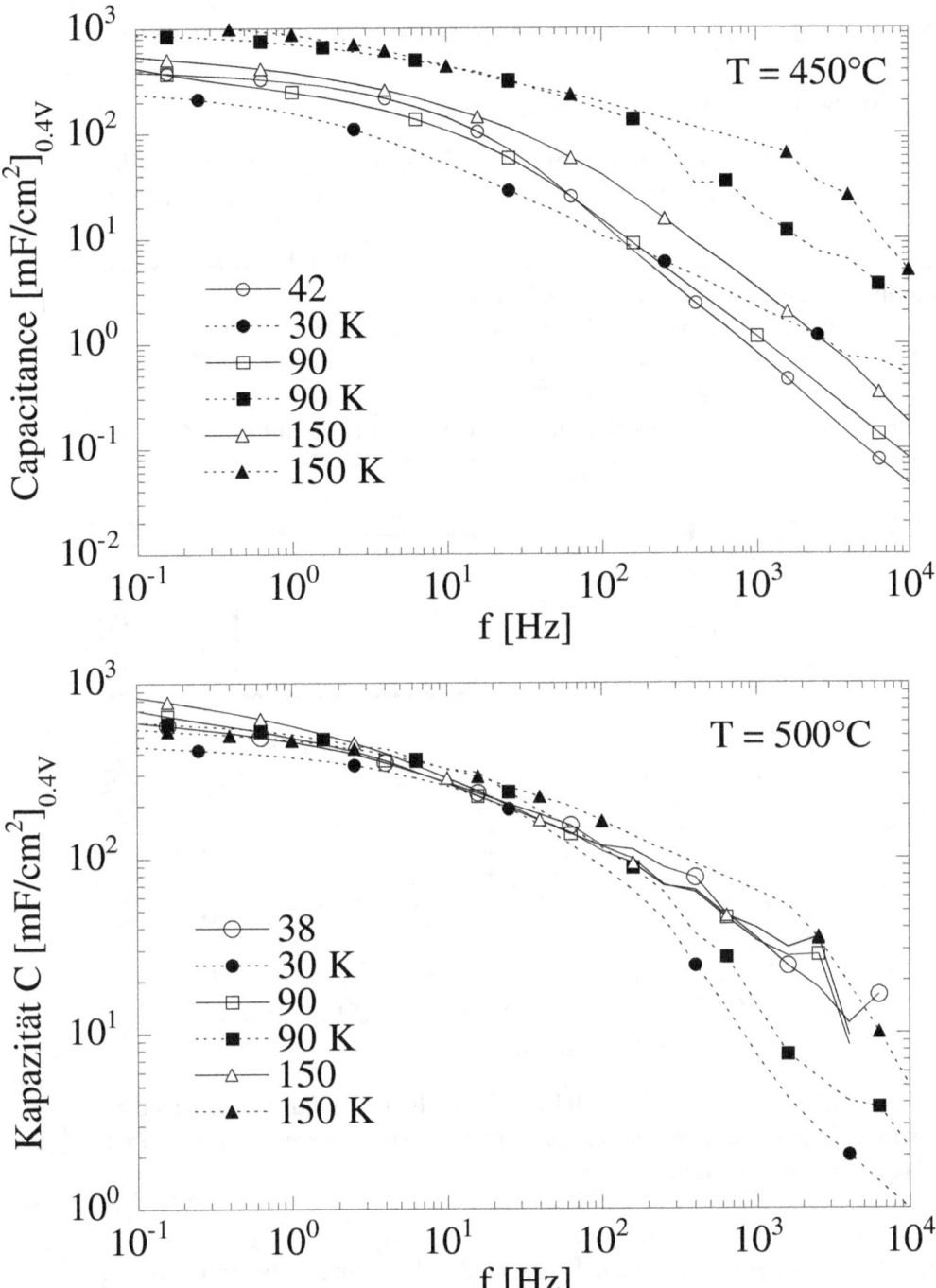

Figure 6.27: Top: Frequency response of the capacitance of K800 GC 1mm, oxidized at 450°C.
Bottom: Frequency response of the capacitance of K800 GC 1mm, oxidized at 500°C.
Activation times in minutes are denoted accordingly in the legend. For comparison, the SIGRADUR®K samples are displayed as well. These samples are marked with a K beside the activation time. Bias potential was 0.4 Volt.

6.3.2 Influence of O_2-Concentration

To clarify whether the concentration of the oxydant gas, O_2, affects the film growth and the capacitance, activation with different gas concentrations were carried out [111] [7].
For these experiments the samples were activated in a so called *Atmos-box*, which was built in the furnace and heated by the furnace.
The gas was supplied as reported in the experimental part of this thesis.
SIGRADUR®K samples were activated at 450°C for various times with oxygen partial pressure of 5%, 20%, 50%, and 100%.
The data for the capacitance are listed in Table 6.5 and plotted in Figure 6.30 for better view.

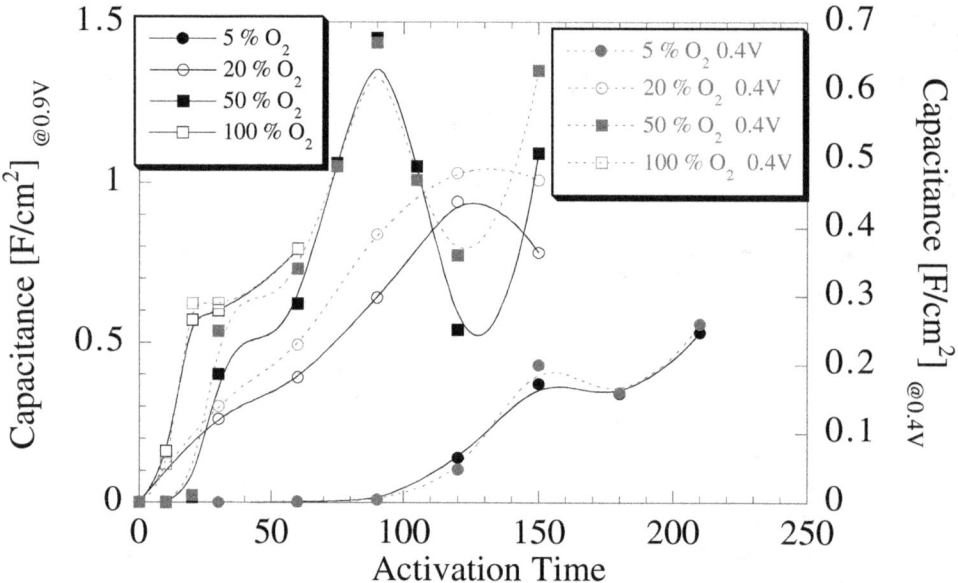

Figure 6.28: Capacitance of SIGRADUR®K with 1 mm thickness, activated at 450°C with various gas concentrations. Left (right) axis concerns data measured at 0.9 V (0.4 V) bias potential.

Figure 6.30 displays the capacitance measured at different bias potentials (0.4 V and 0.9 V).
The values for the capacitance measured at a bias potential of 0.4 V are by a factor of roughly two larger than the capacitance measured at 0.9 V.

---

[7]The experiments were made and the most expressions were derived by Olivier Merlo, Université de Fribourg, during his internship at PSI in summer 1997.

Capacitance [mF/cm^2] for various times and concentrations											
% O$_2$	10	20	30	60	75	90	105	120	150	180	210
5	-	-	5.0 E-5	2.1 E-3	-	8.8 E-3	-	0.14	0.37	0.34	0.53
20	-	-	0.26	0.39	-	0.64	-	0.94	0.78	-	-
50	4.6 E-4	0.015	0.40	0.62	1.06	1.34	0.31	0.31	1.09	-	-
	-	-	-	-	-	0.25	1.05	0.54	-	-	-
	-	-	-	-	-	1.45	-	-	-	-	-
100	0.16	0.57	0.60	0.79	-	-	-	-	-	-	-

Table 6.5: Capacitance of SIGRADUR®K for various gas concentrations. The activation temperature was 450°C. Samples were measured at 0.9 V bias potential.

As already mentioned in the introduction in this section, the capacitance measured at 0.4 V is usually larger than the capacitance measured at 0.9 V due to the surface groups (quinone-/hydroquinone-like, [78, 79, 80, 81, 82, 83]) on the sample.
The plotted data clearly indicate that the profile of capacitance growth is shifted towards smaller activation times, when the O_2 partial pressure is raised.
The saturation capacitance (and saturation film thickness) is reached earlier, when the O_2 pressure is raised.
It is stressed that this observation is directly in line with the predictions of the film growth model established in Chapter 6.2.
However, minor deviations from the capacitance growth experimental data and calculated film thicknesses are observed. These deviations will be studied in the following section (6.3.3).

6.3.3 Capacitance after Weak Activation

For samples activated only weakly (short time activation either at low temperatures or with low concentration of the oxydant), it is observed that the capacitance per sample area increases not convex but concave (see Figures 6.17 and 6.18) [112].
According to the observations made on the film thickness growth, which exhibits a convex behaviour with oxidation time due to the properties of the Generalized LambertW function, the evolution of the capacitance should exhibit the same behaviour.
The Generalized LambertW Function $\mathcal{G}$ is a convex function for its positive arguments (equation 6.46), as also observed for the experimental film thickness data.
When activation proceeds due to longer activation times or higher oxydant concentrations, the curves for the capacitance become convex, and the LambertW behaviour is recovered (see Figures 6.15 and 6.18). Obviously there is a crossover

6.3. INFLUENCE OF ACTIVATION PARAMETERS ON ELECTRODE PERFORMANCE

from the growth of normal reproduction to a logistic growth [113].
The concave part of the curve is particularly well visible for samples either activated with low O_2 partial pressure or at low activation temperature.
One explanation for this effect could be that at the initial stage of activation the pores are too small for the electrolyte to penetrate.
The concave growth is expected for the film growth process with a power law for t smaller than 1 (actually around 1/2 is found due to diffusion limited growth).
The convex growth is accompanied by a power law for t larger than 1.
A derivation will be presented which explains why the capacitance can grow with a law stronger than any power of t - the exponential growth.
The derivation is based on following considerations [8]:
Assumed that the change of the internal surface area at initial stage of reaction is proportional to the weight loss, and the weight loss be proportional to the surface already available, following relations hold:

$$\frac{dA}{dt} \sim -\frac{dm}{dt} = k \cdot c_{O_2} \cdot A , \qquad (6.56)$$

A being the surface area and m being the mass of GC, k being the reaction rate, c_{O_2} being the concentration (i.e.: partial pressure) of oxygen. It follows

$$\frac{dC}{dt} = k^* \cdot c_{O_2} \cdot C \qquad (6.57)$$

C being the capacitance and k^* being also a reaction rate, which differs from k by only a constant factor. Separation of variables C and t and integration yields the solution

$$C(t) = C_0 \cdot exp\left(k^* \cdot c_{O_2} \cdot t\right) \qquad (6.58)$$

with an initial capacitance of C_0, which should be around 20 $\mu F/cm^2$. Therefore, during initial stage of activation the growth follows an exponential.
As the argument of an exponential is dimensionless, the product of the concentration and the constant k is a reaction rate similar to the reaction rate α in Equation 6.11, which describes the burn-off of GC during activation (it must be regarded as a burn-off *velocity*, because it has the dimension length/time).
The exponent was determined from the capacitance at a specific value (0.6 F/cm^2 at 0.4 V bias), as listed below in Table 6.6.

As the logarithm of the ratio of initial capacitance (20 $\mu F/cm^2$ for the flat and smooth, non-activated GC sample) and final capacitance is taken constant for a

---

[8]Part of the considerations were made and part of the expressions were derived by Olivier Merlo, Université de Fribourg, during his internship at PSI in summer 1997.

Capacitance [mF/cm^2] at 0.4 V		
c_{O_2} [%]	Time [min]	Capacitance [F/cm^2]
5	210	0.6
20	90	0.6
50	60	0.6
100	30	0.6

Table 6.6: This data serve to evaluate the reaction rate k. The same value for the capacitance (0.6 F/cm^2) is measured at different concentrations and different activation times.

specific set of activation times and concentrations, the constant k can be determined by following relation:

$$ln\left(\frac{C(t)}{C_0}\right) = k \cdot c_{O_2} \cdot t \qquad (6.59)$$

The data in Table 6.6 were fitted linearly in a plot $1/t$ vs. c with the result that the reaction constant yields $k = 0.00034$/min. This value concerns the reaction at 450°C.

If we assume that the reaction under consideration here (activation, opening of pores) is the same as studied in the film growth model, equation 6.46 has to be corrected as follows:

$$C(t) = C_0 \, exp\left(\frac{\alpha\,t}{\lambda}\right) = C_0 \, exp\left(\frac{b\,k\,c\,t}{\rho\,\lambda}\right) , \qquad (6.60)$$

λ being a characteristic length and α being the burn-off constant in Equation 6.21.

6.3.4 Capacitance of Thermally Activated GC with Different HTT

As the extension of the graphene sheets and crystallite size in GC and also the pore size depend on the pyrolysis temperature (HTT) [16], GC with different HTT was thermally activated and measured.

It is expected that the samples with a higher HTT have a smaller capacitance per sample area, because the film thickness should be smaller, as already known from the structure analysis part (Chapter 5) in this thesis.

GC with a HTT from 800 (K800), 1000 (SIGRADUR®K) and 2200°C (SIGRADUR®G) were purchased from HTW. Results will be presented in this subsection.

Additionally, samples from SIGRADUR®K and K800 were pyrolyzed at PSI up to 1250°C (*post-pyrolysis*) in order to have one more data point in the variation of the temperature parameter.

GC Pyrolyzed at 800°C

As the film thickness of K800 GC samples is expected to be larger than the film thickness of SIGRADUR®K (Section 6.2.4.), the capacitance per sample area also should be larger.

During activation of K800 GC it turned out that the regime at which cracks occur is found at somewhat lower temperatures than for the SIGRADUR®K samples (Section 6.1.1., Figure 6.8). Figure 6.31 displays the increase of capacitance of GC (1 mm thickness, pyrolyzed at 800°C and 1000°C) as a function of activation time for various activation temperatures.

After activation the K800 samples were very rough on the surface. The grade of cracking was more pronounced for the K800 samples than for the SIGRADUR®K samples. Therefore it may be possible that a somewhat higher capacitance is measured for the K800 samples, because electrolyte may creep under the varnish and soak a somewhat larger area than the area which should be exposed to the electrolyte.

The differences in the maximum capacitance of either type of GC are not very large. Error margins were not taken into account for these measurements.

After around 30 minutes of activation at 400°C, K800 reaches a higher capacitance than SIGRADUR®K, but after 150 minutes of activation SIGRADUR®K has a higher capacitance. This also holds for activation at 450°C.

But with 500°C activation temperature the K800 samples have always (all times) a higher capacitance than SIGRADUR®K samples.

Despite a possible slight overestimation of the capacitance of the K800 samples, summarizing one may conclude that K800 GC can be activated in shorter time than SIGRADUR®K.

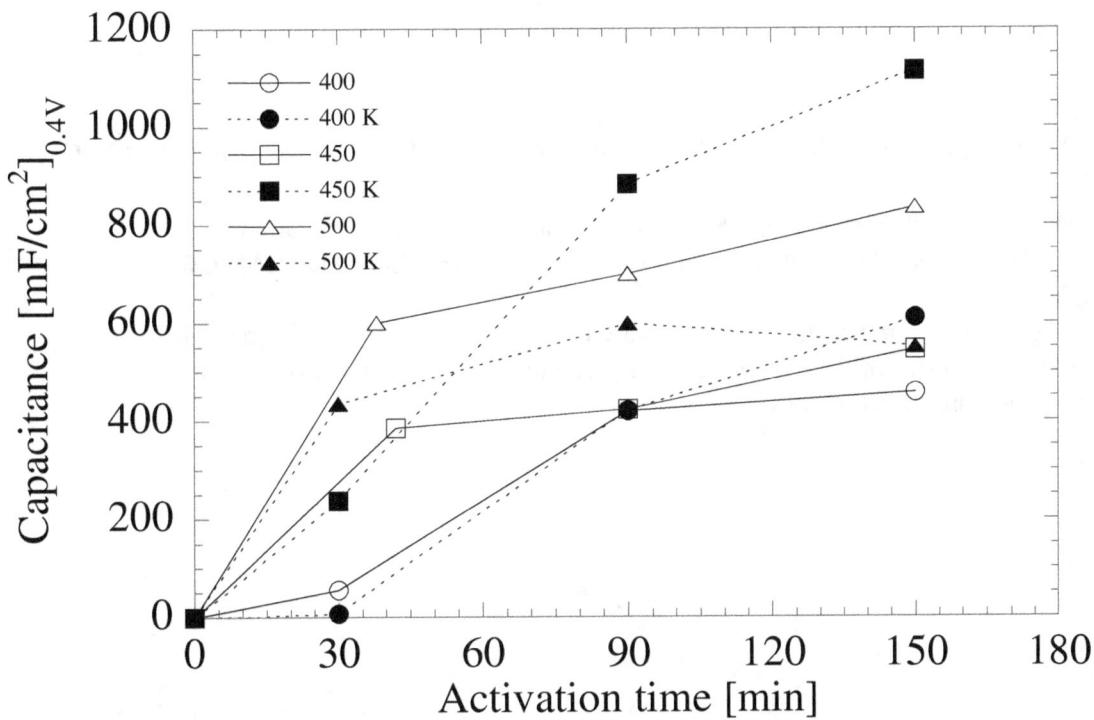

Figure 6.29: Comparison of capacitance of GC K800 and SIGRADUR®K with 1 mm thickness for various activation times and temperatures. Filled symbols concern GC with a HTT of 1000°C, open symbols denote GC with a HTT of 800°C. The solid and dotted lines serve as a guide to the eyes.

6.3. INFLUENCE OF ACTIVATION PARAMETERS ON ELECTRODE PERFORMANCE

Probably the film thickness of the K800 GC sample is larger after activation for a specific time at 400°C and 450°C, than it is in the case of the SIGRADUR®K GC samples.

This explanation is reasonable because the GC pyrolyzed at lower temperatures has a higher density of defects, which is a precondition for diffusion of reactant gases to the unreacted core of GC.

Thermal Activation of SIGRADUR®G

Glassy Carbon pyrolyzed at 2200°C (SIGRADUR®G) cannot be activated thermochemically so well, in contrast to SIGRADUR®K. When the SIGRADUR®G is thermally oxidized, the transversal planes (top and bottom) of the sample are less affected than the frontal and sagittal planes (sides, which were sewed). In the side planes a hollow develops due to a more pronounced burn-off. Obviously the outer skin of SIGRADUR®G is more resistant against activation than the core material.

These findings are in line with results from thermo-gravimetrical analysis (TGA), which are not further reported in this thesis. The oxidation of SIGRADUR®G required higher temperatures than the oxidation of SIGRADUR®K. In particular, burn off is more pronounced at edge planes than on top and bottom of the sample. SIGRADUR®G samples of 1 mm thickness had been thermally activated in a furnace at various temperatures [9], as displayed in Table 6.7.

Muffle Furnace, no air flow		
Temperature °C	Time [min]	Capacitance [$\mu F/cm^2$]
500	30	30
600	30	520
600	180	1400
700	30	55

Tubular Furnace, 20 l/h air flow		
Temperature °C	Time [min]	Capacitance [$\mu F/cm^2$]
550	30	142
600	30	182
650	30	116
725	30	85
750	30	93
800	30	160

Table 6.7: Capacitance of thermochemically oxidized GC SIGRADUR® G, oxidized in air at various temperatures and times. Note that the capacitance values are in μF.

---

[9] Data and experiments by courtesy of Dr. Melanie Sullivan and Dr. Rüdiger Kötz

6.3. INFLUENCE OF ACTIVATION PARAMETERS ON ELECTRODE PERFORMANCE

Attempt was made to remove the skin from SIGRADUR®G with emery paper. When 50 to 100 μm of the GC were removed from bottom and top side of the samples, they were thermally activated at various temperatures and subsequently measured with EIS.

The capacitance and diffusive Resistance of the samples with the removed skin are displayed in Table 6.8. Samples were activated for 40, 80 and 120 minutes

Capacitance [mF/cm^2]					
Time [min.]	450°C	475°C	500°C	550°C	600°C
40	-	-	3.47	-	-
80	2.6	3.74	5.38	0.24	$\rightarrow 0$
120	-	-	0.13	-	-

Diffusive resistance [mΩcm^2]					
Time [min.]	450°C	475°C	500°C	550°C	600°C
40	-	-	41	-	-
80	70	61	68	25	$\rightarrow 0$
120	-	-	$\rightarrow 0$	-	-

Table 6.8: Capacitance and diffusive resistance of thermochemically oxidized GC SIGRADUR®G. The skin of the samples was removed with emery paper.

at temperatures between 450°C and 600°C. The highest value for the capacitance was obtained after 80 minutes activation at 500°C (5.38 mF/cm^2).

For higher oxidation temperatures the capacitance decreases rapidly. This is observed for all types of GC, because the high oxidation temperature causes a higher reaction rate (after Arrhenius by an exponential), therefore a more pronounced burn off of the film and subsequent a smaller saturation film thickness (see Section *Model for Film Growth*).

Probably the active film thickness in thermally oxidized SIGRADUR®G is very small. The film thickness can be roughly estimated as follows:

Assumed a volumetric capacitance of around 100 F/cm^3, which is a reasonable value for state-of-the-art EDLC today, we find a value of 10 mF/cm^2 per 1 μm film thickness. As the maximum value for the capacitance is around 5 mF/cm^2, we conclude that the film has a thickness smaller than 1 micron.

Values for the capacitance and diffusive resistance of samples not sanded prior to oxidation were found to be by a factor of two lower.

The magnitude of the diffusive resistance is similar to those of K 1mm thermally activated, although the film thickness in SIGRADUR®G is smaller than the film thickness in SIGRADUR®K.

The capacitance of activated GC samples, the skin of which was not removed prior to activation, exhibit a capacitance with one order of magnitude lower than those without the skin.

Such samples were also thermally activated one hour in oxygen with concentrations of 50% and 100% [111][10]. Results for the capacitance are displayed in Table 6.9. As already mentioned in previous sections, the capacitance of a smooth sur-

Capacitance at 0.4 V [$\mu F/cm^2$]						
% O_2	500°C	550°C	600°C	650°C	700°C	750°C
50	55	134	132	94	87	93
100	-	-	1633	119	-	-

Capacitance at 0.9 V [$\mu F/cm^2$]						
% O_2	500°C	550°C	600°C	650°C	700°C	750°C
50	81	148	83	102	553	111
100	-	-	790	69	-	-

Table 6.9: Capacitance of thermochemically oxidized GC SIGRADUR® G, oxidized in oxygen for one hour. Note that the capacitance values are in mF.

face is around $20 \mu F/cm^2$.

After one hour of activation at 500°C the capacitance increases only by a factor of around 2 - 3. The highest achieved value for the capacitance after 1 hour of activation is around $500 \mu F/cm^2$, measured at 0.4 V bias potential. This value is not significant, because it was not reported to be reproducible. However, this value can serve as an upper limit for the capacitance obtained for one hour of activation at temperatures between 500 and 750°C in 50% oxygen.

When samples are activated at 100% oxygen concentration, the capacitance increases by nearly a factor of 10.

According to the film growth model presented in Section 6, an increase of the oxygen concentration should accelerate the activation in so far, that the saturation

---

[10] Data and experiments by courtesy of Olivier Merlo.

6.3. INFLUENCE OF ACTIVATION PARAMETERS ON ELECTRODE PERFORMANCE 147

film thickness (and therefore saturation capacitance) is obtained in a shorter time than for a lower concentration of oxygen.

According to equation 6.50, which represents the series expansion for the expression for the film thickness, an increase of the concentration by a factor of two should also increase the film thickness by a factor of $\sqrt{2}$ - at least in the range where the zero-order approximation is valid [11].

This effect cannot explain the increase of capacitance by the factor of 10.

Possibly deeper regions are reached during activation which have another structure than the upper regions (still with the skin), so that after some burn-off of the upper regions material with a larger possible capacitance is obtained.

In contrast, the capacitance of the samples activated at 650°C is lower after oxidation with 100% oxygen concentration. As only one measured value is present for these parameters, data derived here can hardly serve as a proof.

Summarizing one may conclude, however, that the GC pyrolyzed at 2200°C cannot be activated so easily as the GC pyrolyzed at 800°C or 1000°C, because even at high activation temperatures capacitance values of only around 1 mF/cm^2 are obtained.

One reason for this is the presence of a protecting *skin*, which can be removed mechanically so that somewhat larger capacitance values of still less than 10 mF/cm^2 are obtained.

As will be discussed later in this thesis, the defect density in GC plays an important role for activation, and the skin mentioned probably has less defects than necessary for a fast film growth.

Thermally Activated GC Pyrolyzed at 1250°C

The GC commercially available from the manufacturer (HTW GmbH) was pyrolyzed at 800°C, 1000°C and 2200°C. Capacitance measurements (Figures 6.35 and 6.36) reveal that the maximum possible film thickness decreases, when the GC has received a higher pyrolysis temperature.

With 2200°C HTT nearly no capacitance is obtained even at high reaction temperatures, and with 1000°C HTT a much higher capacitance with a larger film thickness is obtained.

However, it remained a little unclear whether there is really a trend between HTT and maximum film thickness, because the samples pyrolyzed at 800°C did not exhibit a significant increase of capacitance.

To clarify this, samples with a HTT at 1250°C were prepared by a post-pyrolysis of K800 GC or SIGRADUR®K. The pyrolysis was carried out in a vacuum furnace ($p \leq 10^{-6}$ mbar) [12] after a pyrolysis schedule as displayed in Figure 6.32.

---
[11] For the activation of SIGRADUR®G this approximation is valid because the films are thin.
[12] Eugen Groth, PSI Dept. of Nuclear Energy and Safety.

These samples will be termed K1250 furtheron in this thesis. It was expected that this treatment alters the structure of the GC significantly.

It is particularly expected that the lateral extension of the graphene layers is increased and that also the stack height of the graphene layers is increased during pyrolysis at a HTT of 1250°C [16]. As a result, the film thickness and maximum obtainable capacitance should be smaller than in the case of the samples pyrolyzed at 800°C and 1000°C.

Additionally, a larger pore size is expected and the frequency response of the capacitance should be altered in so far as the decay of capacitance with increasing frequency should be less drastic than in the case of GC pyrolyzed at lower HTT.

X-ray diffractograms of a 110 micron thick K1250 and its precursor K800 are

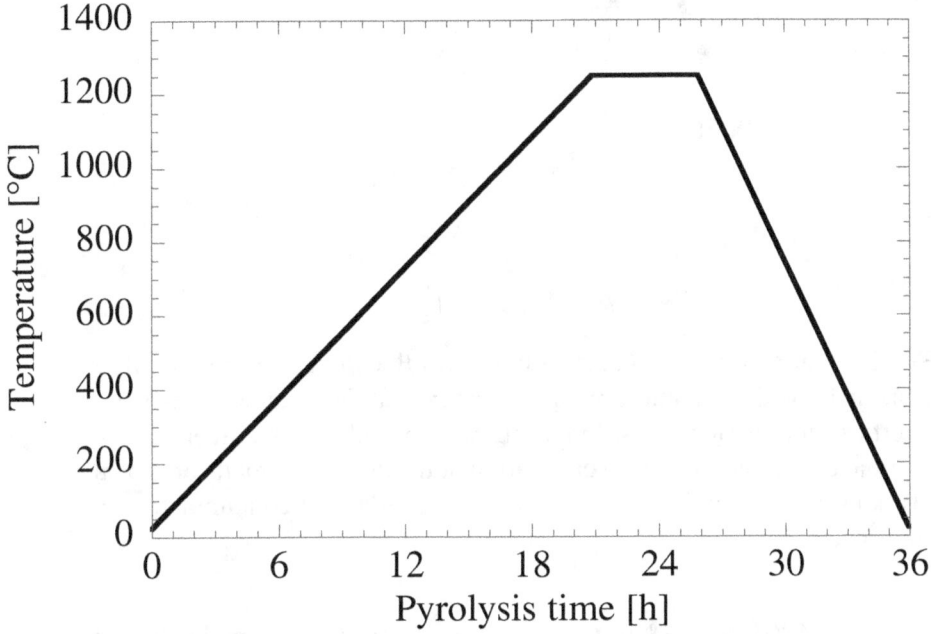

Figure 6.30: Pyrolysis schedule for the K800 GC samples.

displayed in Figure 6.33. First the K800 GC sample was measured, than this sample was after-pyrolyzed as described, denoted K1250 furtheron, and then another XRD diffractogram was measured.

The K1250 sample has a higher (002) peak than the corresponding K800 sample, which reveals a higher structural order in the K1250 samples. Also, the full width at half maximum (FWHM) is smaller than that of the K800 sample, which reveals that the graphene layers are more extended in the K1250 sample. Post-pyrolysis therefore leads to larger (002) diffraction peaks with a smaller FWHM.

The expected decrease of the graphene mean interlayer distance upon annealing

6.3. INFLUENCE OF ACTIVATION PARAMETERS ON ELECTRODE PERFORMANCE

was not observed. Probably a longer pyrolysis time is necessary to change the sample structure so far that this effect can be observed.

Similar behaviour was found for the 1 mm samples. Figure 6.34 displays diffrac-

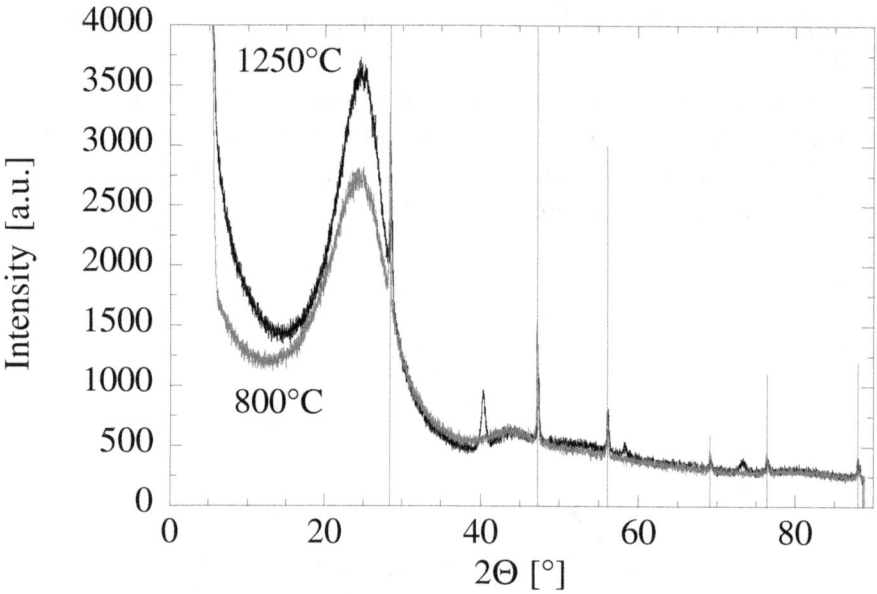

Figure 6.31: Comparison of GC K800 with 110μm thickness, before and after post-pyrolysis at 1250°C. The curve with the larger peak is obtained after post-pyrolysis. Vertical lines indicate position of diffraction peaks from silicon, which was used as a standard material for the calibration of the 2Θ axis. Small additional peaks at positions 40°, 58° and 74° arise from tungsten filament contamination in the vacuum furnace.

tograms of a 1 mm SIGRADUR®K sample and of 1 mm K800 samples before and after post-pyrolysis.

The precursor with 800°C pyrolysis temperature has the lowest intensities for the prominent diffraction peaks in GC, while the SIGRADUR®K and K1250 have systematically larger peaks.

The XRD investigation of GC with different pyrolysis temperatures reveals that the structural order of GC is increased with increasing HTT. Viceversa, the density of defects must decrease. This behaviour of GC is also reported in [16].

To investigate the influence of the HTT on the capacitance, GC samples of 1 mm thickness pyrolyzed at 1000°C and 1250°C were thermally activated for 96 minutes at 450°C.

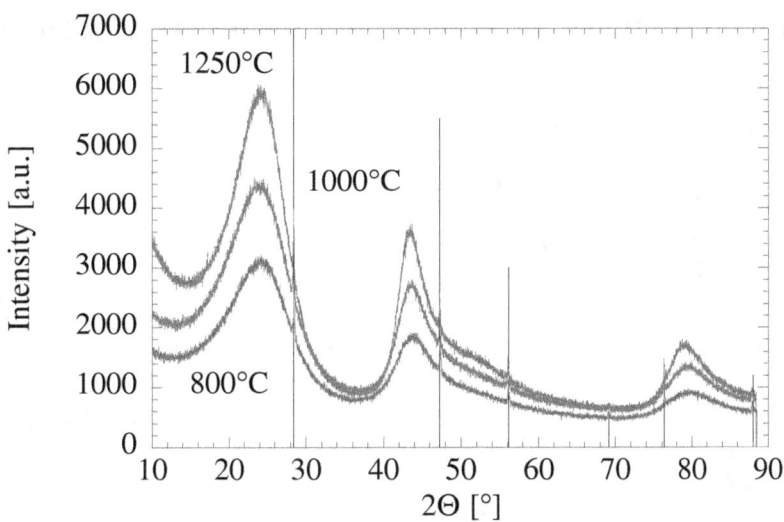

Figure 6.32: Comparison of diffractograms of GC with 1 mm thickness, which were pyrolyzed at 800°C (lower curve), 1000°C and 1250°C (upper curve). The vertical lines denote the peak positions of the silicon reference material.

Figure 6.35 displays the capacitance of the samples as a function of the ac frequency. Samples were measured with EIS and at dc bias voltage of 0.9 V.
Apart from the frequency behaviour, it is obvious that the capacitance at low frequencies (0.1 Hz) is by a factor of 4 smaller for the sample pyrolyzed at 1250°C (48 mF/cm^2) than for the sample pyrolyzed at 1000°C (200 mF/cm^2).
The SIGRADUR®K sample experiences a steep decrease of capacitance at around 2 Hz, while the capacitance of K1250 remains rather stable up to even around 10 Hz.
In the frequency range from 10 Hz to 10^4 Hz no significant change of decrease of the capacitance is observed.
The K1250 sample shows a slight tendency to higher values for the capacitance at frequencies higher than 2000 Hz. But for high frequencies the influence of the inductivity of wires between sample and electrometer may lead to an overestimation of the capacitance, as becomes usually clear for frequencies higher than 10 kHz, in general.

Figure 6.36 displays the capacitance versus the ac frequency for SIGRADUR®K and K1250 GC samples, activated 90 minutes at 500°C. Due to the higher reaction temperature, the samples have at this rather short time of activation a higher capacitance than after activation at only 450°C. The ratio of the capacitance is around 2.5. The probable reason for the smaller capacitance of the activated K1250 com-

6.3. INFLUENCE OF ACTIVATION PARAMETERS ON ELECTRODE PERFORMANCE 151

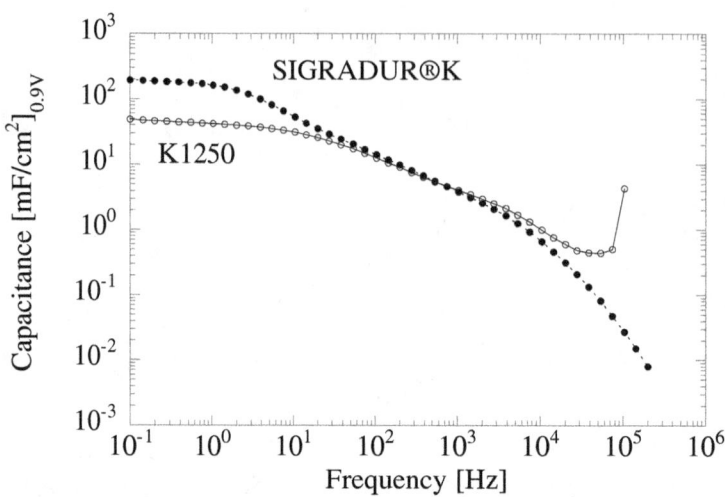

Figure 6.33: Frequency response of the capacitance of SIGRADUR®K and K1250 GC, activated at 450°C.

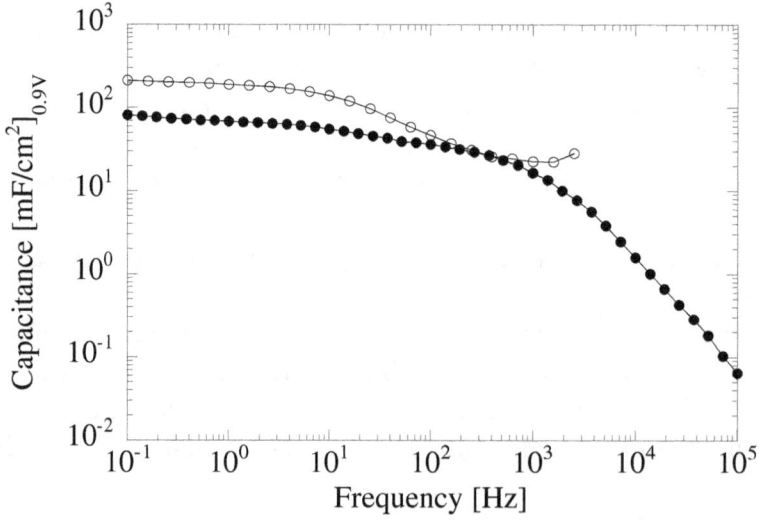

Figure 6.34: Frequency response of the capacitance of SIGRADUR®K (upper curve) and K1250 GC (lower curve), activated at 500°C. Sample thickness was 1 mm.

pared to the SIGRADUR®K sample could be that the film thickness in K1250 is smaller than in SIGRADUR®K.
As the ratio of the capacitance between the two samples at 0.1 Hz is around 2.5, also the ratio of film thickness should be around 2.5. However, for K1250 no direct information on the film thickness was available.

In this section evidence was found that the capacitance of GC is smaller, when it was obtained from activated samples with a higher pyrolysis temperature.
The smaller capacitance is clearly and undoubtedly correlated with the structural differences in the GC.
Low temperature pyrolyzed GC consists of smaller crystallites than high temperature GC. Therefore the defect density is higher in the low temperature GC.
The capacitance per apparent sample area is a reasonably good measure for the active film thickness in activated GC. High temperature GC therefore has a smaller film thickness than low temperature GC.
To finally quantify this, it would be necessary to carry out screening tests with GC with various pyrolysis temperatures, and the screening parameters would be the activation time and activation temperature.
For every sort of GC, information on the maximum film thickness for a specific optimum activation temperature and a reasonable activation time would be desirable.
The overall maximum film thickness is very probably obtained at some specific optimum activation temperature (SOAT). For instance, for the SIGRADUR®K GC this temperature is around 450°C, for the K800 GC this SOAT is lower than 450°C, probably around 400°C. The SIGRADUR®G with 2200°C pyrolysis temperature has this SOAT at around 600°C. For the K1250 GC, the SOAT is beyond 450°C, because the samples activated at 500°C have a larger capacitance than those activated at 450°C. The SOAT of K1250 is therefore possibly around 500°C.
The relationship between this SOAT, which yields highest capacitance per sample area and maximum film thickness, is plotted versus the pyrolysis temperature in Figure 6.37.

The SOAT obviously depends on the pyrolysis temperature of the GC. When the GC has received a high pyrolysis temperature, also the SOAT is higher. With increasing pyrolysis temperature, the activation temperature to obtain the highest capacitance and film thickness has to be raised.
It is expected that the overall maximum film thickness for various sorts (which is obtained by activation at the SOAT) of GC mainly depends on structural data such as crystallite size and defect density.
One probable scenario for the film growth could be as follows:
The defect density very probably determines the effective diffusion coefficient

6.3. INFLUENCE OF ACTIVATION PARAMETERS ON ELECTRODE PERFORMANCE 153

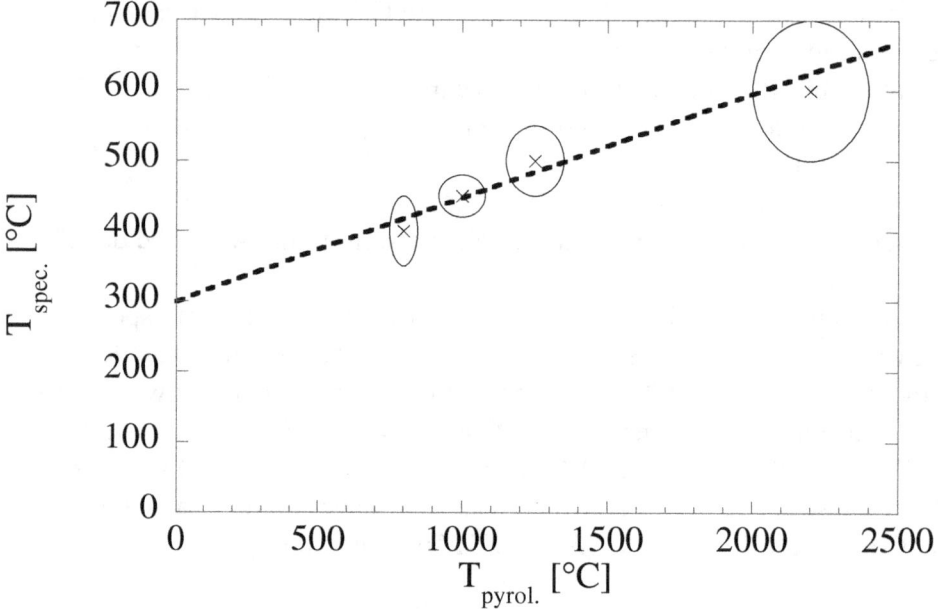

Figure 6.35: The SOAT of activation for the maximum capacitance and film thickness versus the pyrolysis temperature of the according GC. The circles around the measured data (×) denote the uncertainties in the temperatures (error bars). The straight line was obtained from a weighted linear least square fit, but should serve more as a guide to the eyes, because no reason was known that the trend should be linear.

and therefore finally the film growth properties. While GC with small crystallites must have many defects (in majority the grain boundaries), the GC pyrolyzed at higher temperatures has a higher structural order, therefore larger crystallites and less defects.

A volume filled with few large crystallites has less defects than a volume with many small crystallites. The defects possibly serve as nuclei for diffusion paths of reactant gases. Low temperature GC therefore has more diffusion paths which facilitate the film growth so that a thicker film is obtained.

Figure 6.36: Comparison of the capacitance of regular K100 μm (o) and Capton K 150μm (•) GC, measured at 0.4 V bias voltage. The samples were activated at 450°C.

capacitance of the SIGRADUR®K with 100 microns thickness. Probably the Capton GC has a larger film thickness than the SIGRADUR®K with 100 microns thickness. Due to the limited amount of Capton GC material, film thickness determination was not possible.
As displayed in Figure 6.23, the diffusion resistance of Capton GC increases, as

---

[13]Pyrolysis of foils has the benefit that no curing process is necessary. Otherwise, the curing process represents an additional step during production of GC. This raises production costs.

activation time (and therefore film thickness) increases. This behaviour is expected, because diffusive resistance depends on the active film thickness.

Figure 6.39 displays the frequency response of the capacitance of the GC made from Capton.

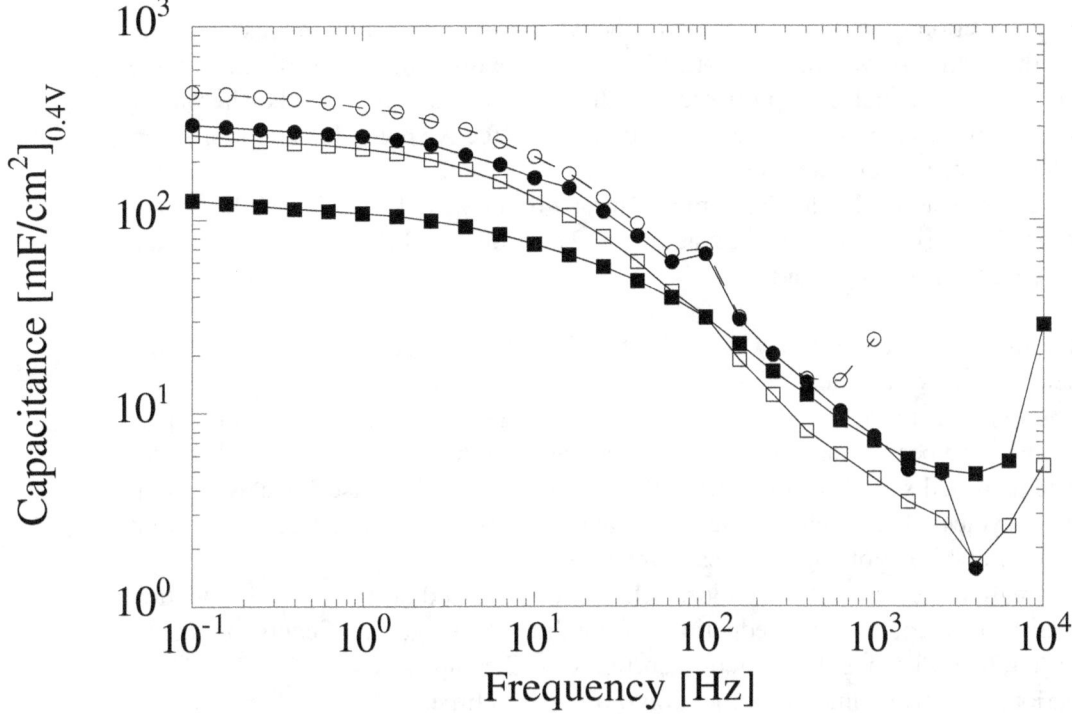

Figure 6.37: Frequency dispersion of the capacitance of thermally activated (450°C) Capton GC. Activation times are marked with symbols: ■ 30 min, □ 60 min, • 90 min, ○ 120 min.

6.3. INFLUENCE OF ACTIVATION PARAMETERS ON ELECTRODE PERFORMANCE

6.3.6 Impact of Reduction on Electrode Performance

Electrochemically activated GC samples exhibit a large capacitance only when they are reduced after activation [13] [14]. This will be discussed in more detail in chapter 6.4.

The reduction is necessary because the activation is an oxidation process with the result that the active film is decorated with surface functional groups which make the film having dielectric properties, i.e. having a low conductivity. Furtheron the surface groups may furnish the pores so that the effective pore diameter is smaller, when surface groups are present.

An electrochemical reduction removes these groups with the result that the capacitance increases by 3 orders of magnitude. The reduction has to be performed at a potential between -0.3 and -0.5 V.

Within this thesis work it was to clarify whether a reduction could improve the performance of the electrodes.

Therefore the thermally activated GC samples were electrochemically reduced after they had been measured for their capacitance in the un-reduced state.

The activated samples were cycled (CV) several times (at least 5 cycles, and up to 30 cycles), and for each cycle the capacitance was determined from the current density at +0.9 V on the back scan direction.

Thereafter, the samples were electrochemically reduced at -0.3 V. The reduction was interrupted, when the reduction current did not change significantly anymore. Figure 6.40 displays the reduction current of GC samples activated at $350°C$ for various activation times, as a function of reduction time.

Apart from the sample activated only 30 minutes, the steady-state reduction current (current for large times t) varies with the activation time.

Samples activated longer time have a larger film thickness, therefore a larger double layer area and therefore a larger reduction current.

After reduction, the samples were cycled again in order to determine the capacitance.

The current density of the first cycle of the activated samples usually was a little higher than that of the reduced samples.

The current density and therefore the capacitance increased similar to a squarerrot-like behaviour with the cycle number. This effect should be related, however, more with the time passed during cycling and not so much on the cycle number.

When after a specific number of cycles a steady state for the current density was obtained, no significant differences ($\leq 10\%$) in the current density and capacitance

---

[14] This is not valid for all types of GC. Electrochemically oxidized GC of SIGRADUR®G with 60 microns thickness exhibited contradictory results. After reduction, the oxidized samples had a lower capacitance and also a lower internal surface area [114].

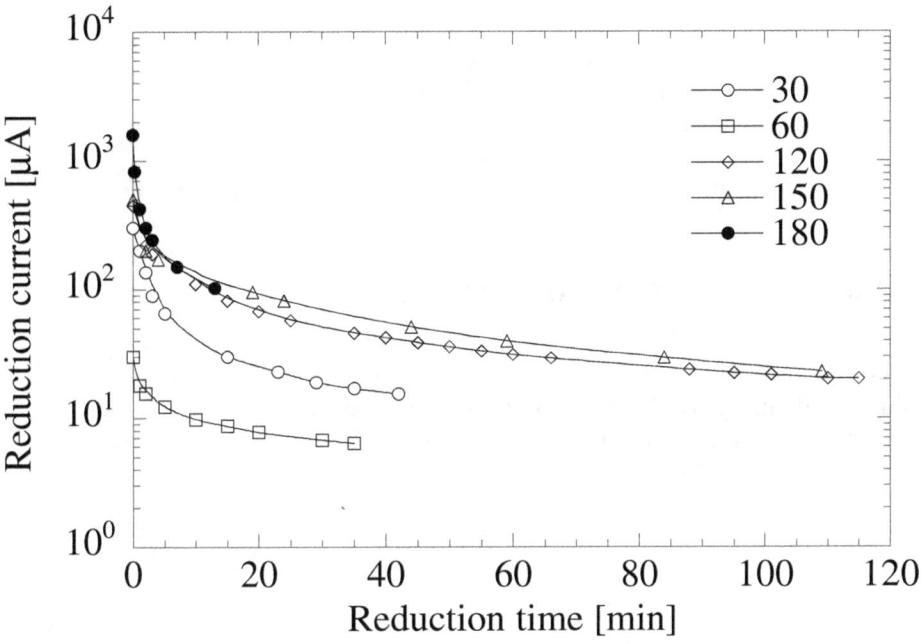

Figure 6.38: Reduction current of activated GC samples (SIGRADUR®K, 350°C, 1 mm thickness). Reduction potential was -0.3 Volt. The electrode surface was around 1 mm² for each sample.

6.3. INFLUENCE OF ACTIVATION PARAMETERS ON ELECTRODE PERFORMANCE 159

were measured.
This is a valuable advantage of thermal activation, because, with respect to a potential industrial application, the reduction step is obsolete.

6.3.7 Estimation of Capacitance by Geometrical Considerations

a) There follows a simplistic estimation of the maximum amount of electric charge that can be deposited on a surface. Screening effects by anions and hydrated ions and interactions between surface and ions were not taken into account. Consider an ideally smooth crystal surface and charge carriers (ions) having a size which is similar to the size of a carbon atom. The density of atoms on a surface plane of SIGRADUR®K (ρ=1.53 grams/cm^3, mass number 12) is σ=1.78·10^{15} atoms/cm^2 - the mean atomic distance of which yields around 2.37 Å. Any interstitial site can be filled with one charge carrier (ion).
With the assumption that each ion carries at least one elementary charge of $e = 1.602 \cdot 10^{-19} C$ at a voltage of 1 V, a minimum capacitance of $286 \mu F/cm^2$ can be assigned to this surface area.
This value is by a factor of 14 higher than the usually reported experimental value of $20 \mu F/cm^2$ [115], which was experimentally verified also in this laboratory ($23 \mu F/cm^2$) on untreated flat SIGRADUR®K and SIGRADUR®G samples.
In above calculation the repulsive interaction between charges of even sign was neglected. The potential between two elementary charges with a distance of 2.5 Å is known to be 6 V (Coulomb's law). At 1 V potential only, the distance increases to 14.4 Å. Taking the repulsion into account, the number density of ions per surface area decreases by a factor of (14.4 Å/2.37 Å)^2 = 36. The capacitance corrected for the repulsion forces then yields around $286 \mu Fcm^{-2}/36 = 8 \mu Fcm^{-2}$.

b) The capacitance of the samples was only related to their apparent surface area (geometrical surface). However, the internal surface area is quite larger than this. There follows another estimation of the capacitance per internal surface area. Relating the capacitance per apparent surface area to the internal surface area per film volume, the capacitance should be in the range of $20 \mu Fcm^{-2}$.
Consider a sample of SIGRADUR®K, two hours activated at 450°C. The internal surface area of the film material of the sample (BET measurements) and the film thickness (SEM) are known. The active film thickness is around 45 μm (= 45·10^{-4}cm). A sample with an apparent area of 1 cm^2 then has an active pore volume of around 45·10^{-4}cm^3.
The internal surface area in this volume has to be determined yet.
The BET internal surface area of the active film material was determined to be around 1000 m^2/g. It is necessary to determine the mass density of the active film material (m=ρV) in order to know the internal surface area per film volume.
Due to the oxidation process during activation the GC film density is smaller than the GC bulk density.
Starting with the bulk density ρ_{bulk}=1.53 g/cm^3, a mass of m =1.53×45·10^{-4}g = 6.885 mg is obtained. The weight loss during activation at 450 °C is around

6.3. INFLUENCE OF ACTIVATION PARAMETERS ON ELECTRODE PERFORMANCE

1.573 mg cm^{-2}h^{-1}. In two hours, the weight loss is therefore 3.15 mg. The remaining mass of the film is therefore 6.885 mg - 3.15 mg = 3.735 mg.
Film material with this mass has a BET internal surface area of around 1000 m^2/g. So the internal surface area is O=3.735·10^{-3}×1000 m^2 = 3.735 m^2. The sample with 1 cm^2 geometric area has an internal surface area of around 37350 cm^2. The capacitance of this sample is around 1 Farad.
Hence the capacitance per surface area yields 1 F/3.735·10^4 cm^2 or 27 μFcm^{-2}, respectively. This calculated value is in good agreement with the experimentally verified value of around 20 μF/cm^2.

6.3.8 Influence of Electrolyte Temperature on Capacitance and Resistance

It is of considerable interest to know the influence of the temperature on the capacitor performance.
When the capacitor is being utilized for instance in automotive vehicles or portable devices, the surrounding temperature may vary between -30°C and 80°C, depending on real live situations.
In the vicinity of engines even more than 100°C may be possible.
To check for the influence of temperature, the electrolyte temperature T_E in single electrode measurements was varied between -20°C and +65°C. The T_E was changed and kept stable using a cryostat (LAUDA) filled with ethyleneglycol. Temperature was measured with a mercury thermometer.
As an electrode, a piece of GC (SIGRADUR®K, 1mm), activated for 90 minutes at 500°C, was used.
Fig. 6.41 displays the capacitance [15] and the diffusive resistance [16]. The capac-

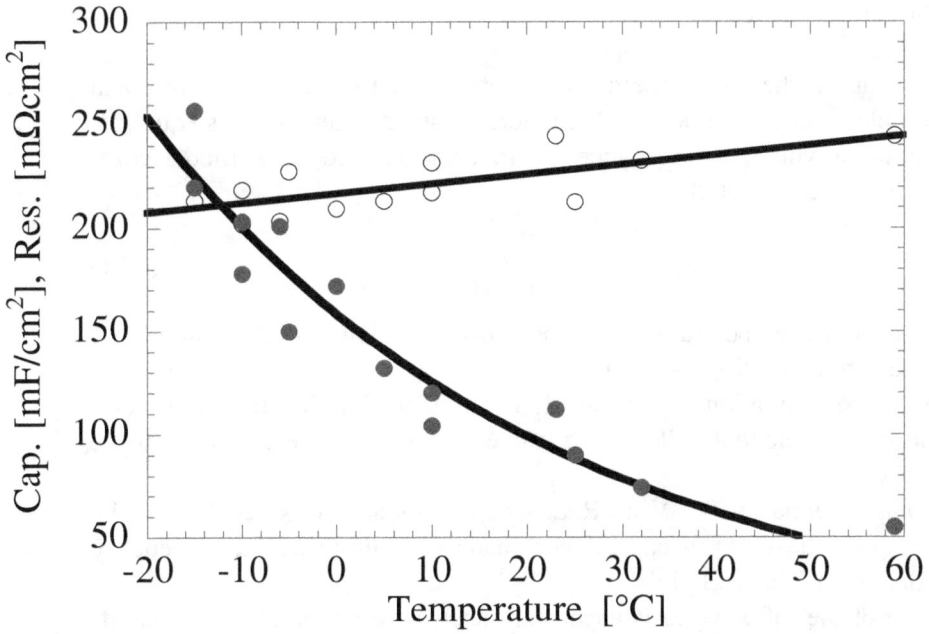

Figure 6.39: Capacitance (○) and diffusive resistance (●) of SIGRADUR®K with 1 mm thickness as a function of the electrolyte temperature. Drawn lines act as guides to the eyes. The sample was activated 90 minutes at 500°C.

---

[15] measured at 0.9 V bias voltage and 0.1 Hz.
[16] determined from the high frequency intercept.

itance exhibits a slight increase with increasing T_E. The diffusive resistance is increasing slightly, when electrolyte temperature T_E is lowered. For increasing T_E it is observed that the low frequency part is not a straight line anymore, as expected for a real capacitor.

The impedance curve bends slightly over to a semi circle, which gives evidence that an electrochemical reaction takes place in the system.

Therefore at 0.1 Hz, at which frequency the imaginary part is measured for the determination of the capacitance, the capacitance is slightly overestimated by an amount of around 0.5 $mFcm^{-2}K^{-1}$. The origin of this additional capacity is unclear.

It is possible that chemical reactions occur in the cell environment (brass sample holder) which cannot be avoided easily with the experimental setup used here and which may not be attributed to the GC system under investigation.

During measurement it turned out that the radius of the semi circle sensitively depended on the accurateness of the measurement and the cleanliness of the sample environment.

At elevated temperatures, vapour from the electrolyte wetted all the cell interior and also the sample holder.

When T_E is raised, the electrochemical potential at the sample does not equal the standard electrochemical potential anymore. Instead, after Nernst's equation, those potentials at which the impedance spectra are measured, are shifted accordingly and have to be corrected.

$$\phi_0 = \phi_{00} + \frac{RT}{zF} \ln \frac{c_{ox}}{c_{red}} \qquad (6.61)$$

T being the absolute temperature, R the gas constant (8.314 J/(mol K)) and F being the Faraday number (96400 As/mol).

Assuming that concentrations of species do not change significantly during temperature raise, the potential shift over a range of 50 K can be estimated to be around 10 mV.

As at the bias potential of 0.9 V no Redox capacities are measured (see cyclic voltammograms), the error in capacitance made by measuring the impedance spectra at 0.9±0.01 V is negligible.

The change of the diffusive resistance with electrolyte temperature T_E has its origin in the change of the electrolyte resistance. This follows from the following considerations: The electrolyte resistance was determined from the intercept of the impedance curve (Nyquist plot) with the real axis.

The distance from the origin to this intercept is a resistance caused by the electrolyte resistance and other resistances such as sample holder rod, electric wires to the electrometer, and sample itself.

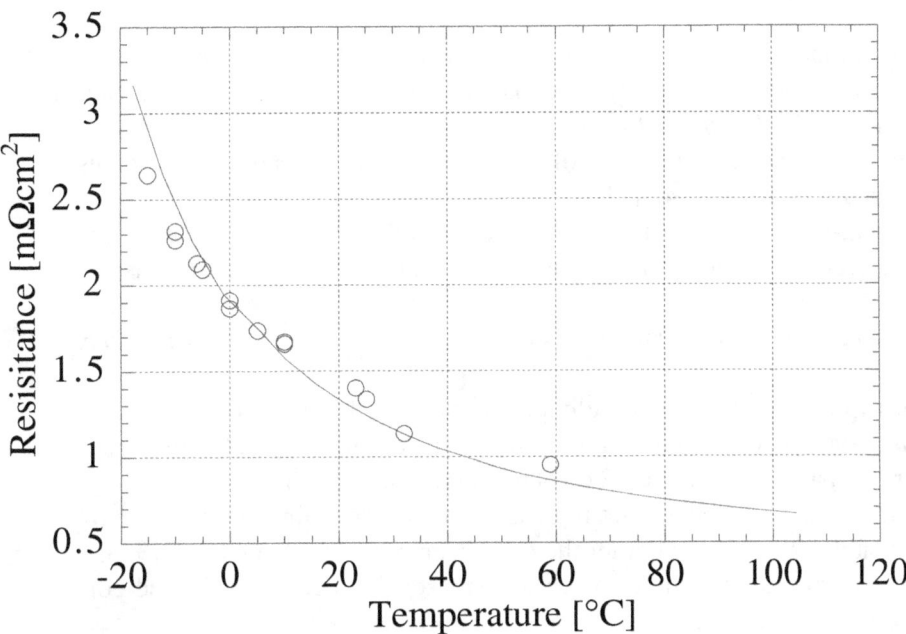

Figure 6.40: Specific resistance of 3 molar sulfuric acid depending on temperature [116, 117].

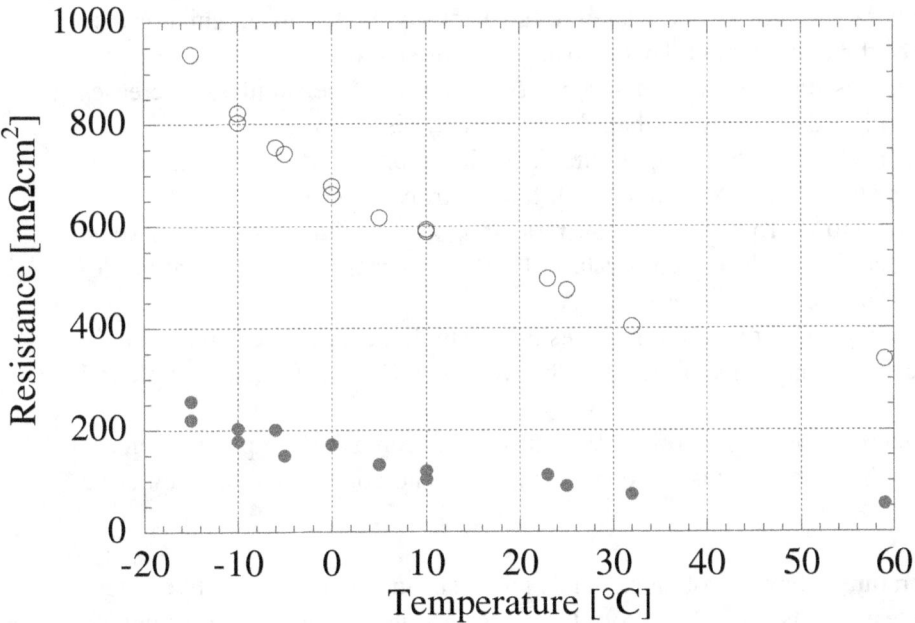

Figure 6.41: Comparison of diffusive resistance (●) and electrolyte resistance (○).

6.3. INFLUENCE OF ACTIVATION PARAMETERS ON ELECTRODE PERFORMANCE

The major resistance contribution is caused by the electrolyte, however, because its conductivity is much lower (0.74 S/cm) than those of glassy carbon (2×10^3 S/cm) and copper wire (6.5×10^5 S/cm, [118]).

Our experimentally determined electrolyte resistance is compared with values found in literature in Figure 6.42 [116, 117].

A direct comparison of our experimental values with the values from literature is not possible, because the electrolyte volume allowing for the ionic current is not exactly known.

Therefore a specific resistance cannot be determined (key word: *Wagner number*, [11]).

However, it is possible to make an indirect comparison of both resistances when one of both curves is scaled with an appropriate factor and then both curves are overlaid for comparison. Figure 6.42 displays such a comparison.

The values from literature (S/cm) are represented by a drawn line and overlaid by our experimental data [$m\Omega cm^2$], after they had been scaled with a factor 1/355.

Measured points and drawn curve match with reasonable accuracy. So the correctness of our data is confirmed.

Figure 6.43 displays the diffusive resistance and the electrolyte resistance depending on T_E. Obviously the electrolyte resistance is increasing more significantly with the electrolyte temperature, than the diffusive resistance.

To make a fair justification of the diffusive resistance as a function of the electrolyte temperature, it has to be related to the electrolyte resistance, which is also a function of the temperature. This is made in Figure 6.44.

The diffusive resistance corrected so far (R_{Diff}/R_{Elec}, dimensionless) increases slightly linearly (-0.00126/K), when the temperature is lowered.

Therefore it is clear that the temperature dependence of the diffusive resistance is a result of the temperature dependence of the electrolyte resistance.

The slight deviation from the constant ratio R_{Diff}/R_{Elec} has unknown reasons. Some explanations could be temperature dependent adsorption, temperature dependent materials electronic resistance, etc.

For a technological application these results mean that capacitance and R_{Diff} in the technical relevant temperature range between -15°C bis +60°C are not altered significantly.

Solely the increase of the electrolyte resistance with decreasing temperature presents a problem, which however is a global feature of ionic conductors and cannot be avoided.

The temperature behaviour of capacitance and resistance is in quantitative agreement with test results on EDLC from the Swiss Montena Corporation (*montena components sa*) [119].

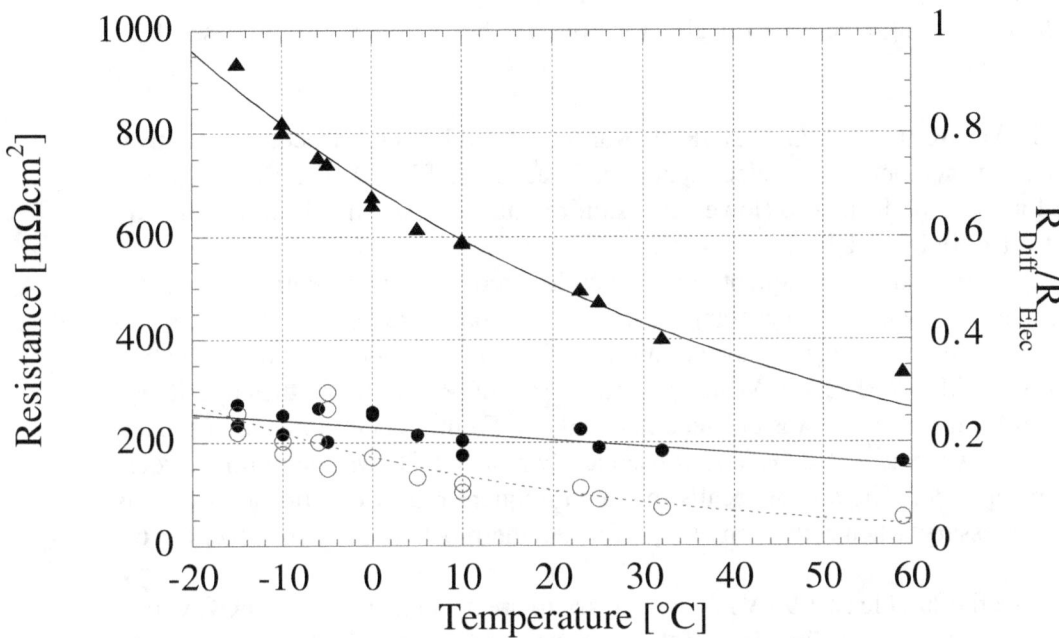

Figure 6.42: Comparison of electrolyte resistance and R_{Diff}. Note there are different axes on left and right abscissae. The lower curve with open circles denotes the measured diffusive resistance. The upper curve with closed triangles concerns the measured electrolyte resistance. The ratio of both curves, R_{Diff}/R_{Elec}, is shown by the curve with the filled circles. The drawn line was obtained from a linear least square fit: $R_{Diff}/R_{Elec}=0.23-0.00126/\Theta$.

6.4 Comparison with Electrochemically Activated Glassy Carbon

6.4.1 General Remarks

Glassy carbon can also be oxidized electrochemically [28, 13] (anodic oxidation). As the scope of this thesis is the thermochemical gas phase oxidation of GC, results concerning electrochemical activation will be reported and discussed here only briefly.

SIGRADUR®K and SIGRADUR®G was tested at PSI for electrochemical activation. It turned out that the electrochemically activated SIGRADUR®G electrodes had a better performance (lower diffusion resistance) than the electrochemically activated SIGRADUR®K.
After activation, a subsequent electrochemical reduction at potentials between -0.3 and -0.5 Volt was necessary, however, to obtain a reasonably large capacitance. A volumetric capacitance of around 80 F/cm^3 was obtained after reduction for SIGRADUR®K at 0.9 V bias potential. For SIGRADUR®G, the capacitance at 0.9 V bias potential was between 200 and 250 F/cm^3. The difference in capacitance between activated GC and subsequently reduced GC was up to three orders of magnitude. The author recalls that a subsequent reduction after activation is not necessary, when activation was performed thermochemically (Chapter 6.3.6).

A potential of at least 1.95 Volt is necessary to create an active film on GC within a reasonable time. An increase of the potential from 1.95 to 2.07 Volt accelerates the film growth by a factor of around 2.5 for SIGRADUR®K and SIGRADUR®G. When SIGRADUR®K is activated instead of SIGRADUR®G, the growth rate increases by a factor of two.
Note that the growth rates are constant, which states that the film thickness by electrochemical activation is a linear function with activation time, while for the thermal activation a more complex relation was found.
  The growth rates for electrochemical oxidation are by far lower than those for thermal activation. While after 2 hours of thermal activation at 450°C a film thickness of around 50 μm is obtained, the film thickness after electrochemical activation does not exceed 5 μm.
Activation potentials higher than 2.1 Volt cause a degradation of the electrode with the result that the resistance of the electrode is unacceptably high.

The growth rates are determined by evaluation of the film thickness of samples activated for various times. Previously this was performed with ellipsometry [13, 43]

Film Growth Rates [μm/h]		
GC-type	1.95 V	2.07 V
K	1.8	4.5
G	0.65	1.63±0.30

Table 6.10: Active film growth rates for electrochemical activation.

and with atomic force microscopy (AFM) [120].
A good and quick estimation of the film thickness in this thesis work was as follows:
When the activated sample is removed from the electrolyte and then dried at ambient atmosphere, the active film cracks off and leaves behind a hollow. With an optical microscope the depth can be estimated with an accuracy of 1 μm by focussing the sample top and the bottom of the hollow. The height difference for both focus is assumed to be the active film thickness [17]. A linear correlation between film thickness and capacitance was found. The largest film (SIGRADUR®K with 1 mm thickness) measured was 18 μm after 4 hours of anodization at 2.07 Volt. The growth rates obtained by this microscopy method are in agreement with the data found by ellipsometry and AFM. With ellipsometry only thin ($\leq 1\mu$m) reduced active films can be measured. But even thick ($\geq 1\mu$m) reduced films could be measured with the optical microscope as described above.

To confirm that no significant amount of film material remained in the hollow after cracking, one sample was measured again to check for the capacitance. The capacitance measured with the empty hollow was around 1% of the capacitance of the sample measured prior to removal from the electrolyte.
The activated SIGRADUR®K film material, which cracked off after drying, was measured with BET to obtain information on the internal surface area. An internal surface area of 3 m^2/g was found, which is a rather low value compared with thermochemically activated GC-powder (see Appendix A) and monolithic samples.

Considering the data in Table 6.10 it is obvious that SIGRADUR®K has a larger film growth rate than SIGRADUR®G.
The capacitance of both types of activated GC differs by a factor of roughly three.

---

[17]It is assumed that the film is only inside the sample. Using AFM [120] it was found that part of the film ($\approx$30%) grows outside the sample, while the other part of the film is in the sample. However, this effect was measured on thin films ($\leq 5\mu$m) only and therefore neglected here.

6.4. COMPARISON WITH ELECTROCHEMICALLY ACTIVATED GLASSY CARBON

The resistance of the electrode from SIGRADUR®K is by three orders of magnitude larger than in the case of SIGRADUR®G. For the SIGRADUR®K a specific diffusive resistance of around 30 kΩcm was measured. This is probably due to a degradation of the SIGRADUR®K during electrochemical activation. When SIGRADUR®K samples are reduced at -0.5 V instead of -0.3 V, the diffusion resistance decreases by a factor of three.

The impedance spectra of electrochemically activated SIGRADUR®K look similar to spectra of powder electrodes (from thermally activated SIGRADUR®K powder).

Obviously the consistency of the GC matrix from SIGRADUR®K suffers from the electrochemical activation more than by thermal activation.

Reduction and Oxygen Content

Table 6.11 lists the oxygen content [18] on 30 minutes thermally activated SIGRADUR®K (1mm thickness) as a function of activation temperature. Data were obtained with X-ray Photoelectron Spectroscopy. With increasing activation temperature the oxygen content also increases.

Table 6.12 displays the oxygen content of SIGRADUR®K and SIGRADUR®G prior to and after electrochemical treatment. The SIGRADUR®K has always a larger amount of oxygen on its surface. In the case of non-activated and reduced GC, SIGRADUR®K has an oxygen content which exceeds the oxygen content of SIGRADUR®G by a factor of two.

From Tables 6.11 and 6.12 it is obvious that thermally activated GC has a lower content of oxygen than electrochemically activated GC. Only after reduction the activated SIGRADUR®G has the same amount of oxygen on its surface as the thermally activated SIGRADUR®K.

6.4.2 Electrochemically Treated Thin GC Sheets

It was found that the subsequent reduction does not in every case yield a higher capacitance. In the case of thin GC sheets (SIGRADUR®G with 60 microns nominal thickness), the capacitance after reduction was smaller than after activation. These samples exhibited a rapid degradation of electrode material during electrochemical activation.

Electrochemically oxidized and subsequently reduced Glassy Carbon (GC) sheets with 60 microns nominal thickness were also investigated with Small Angle X-ray and Neutron Scattering.

---

[18] X-ray Photoelectron Spectroscopy experiments were carried out by Dr. Bernhard Schnyder, PSI.

Thermal Activation	
Temperature [°C]	At% O [K-Type]
400	6.5
450	7.7
475	9.5
500	11.8

Table 6.11: Oxygen content on SIGRADUR®K after 30 minutes thermal activation at various temperatures.

Electrochemical Activation		
State	G-Type [At% O]	K-Type [At% O]
non-activated	3.0	6.5
activated	28.4	30.2
reduced	11.2	22.6

Table 6.12: Oxygen content on GC after 10 minutes electrochemical activation at 1.95 Volt and 5 minutes reduction at -0.3 Volt.

The samples were electrochemically activated at 2.07 Volt and electrochemically reduced at -0.5 Volt. Activation time was 5 minutes and 1 hour, respectively. Reduction time was 15 minutes.

It was found that the internal surface area of the oxidized GC is larger than the internal surface area of GC reduced after oxidation.

It was also found that oxidized GC has a larger capacitance than GC which was reduced after oxidation, which is in contradiction to results previously reported on electrochemically oxidized and reduced GC disks with 1 mm thickness [13, 43].

Figure 6.45 displays cyclic voltamograms of the non-activated and electrochemically activated and reduced sample. All data concern the same sample, but at different stages of treatment. The current density of the electrochemically treated samples is by three order of magnitudes larger than the current density of the non-activated sample. The reduced sample exhibits a smaller current density than the activated sample. These findings qualitatively reveal that oxidation creates accessible internal surface. The capacitance measured with EIS at 0.1 Hz was 36.5 mF/cm^2 after 5 minutes of oxidation and 22.2 mF/cm^2 after subsequent reduction. The ratio of capacitance is 1.64.

6.4. COMPARISON WITH ELECTROCHEMICALLY ACTIVATED GLASSY CARBON

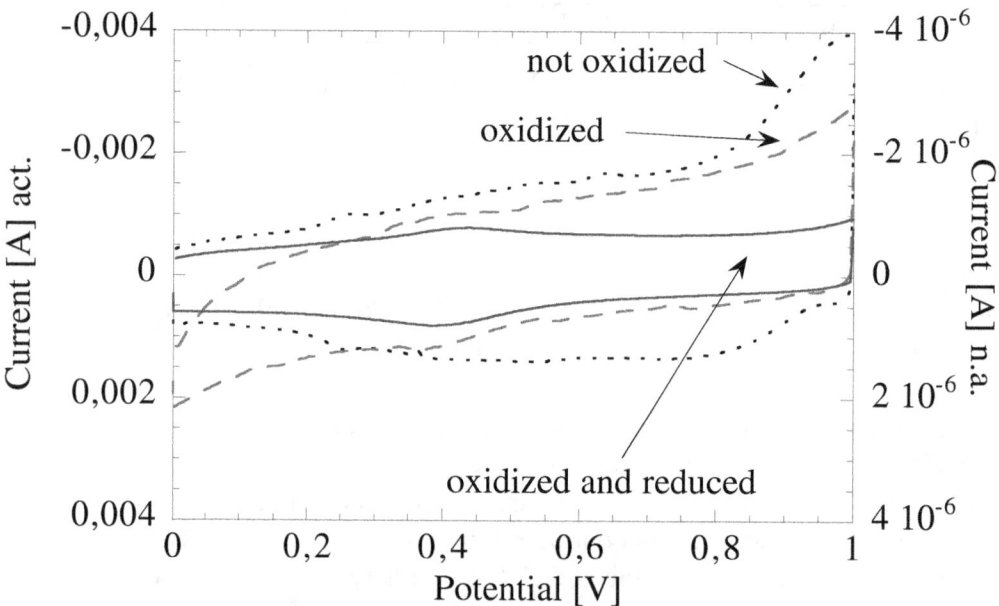

Figure 6.43: Cyclic voltamograms of non-activated and electrochemically activated (5 minutes) and reduced (15 minutes) SIGRADUR®G with 60 μm thickness. Note the two different abscissae on leftside and right side.

The capacitance at 0.1 Hz was around 440 mF/cm^2 for the 1 hour oxidized and 236 mF/cm^2 for the subsequently reduced sample. The ratio of the capacitance is 1.86.
It is recalled that non-activated GC has a capacitance of only 20 μF/cm2.

The dried samples were measured ex-situ with SANS in order to get information on the internal surface area. Figure 6.46 displays the Porod plots of the SANS curves of the samples concerning 1 hour of activation. The SANS data of the sample activated only 5 minutes could not be evaluated, because they differed not so much from the SANS curve of the non-treated sample.

The extrapolated curves to large Q values yield the Porod constants, which are a

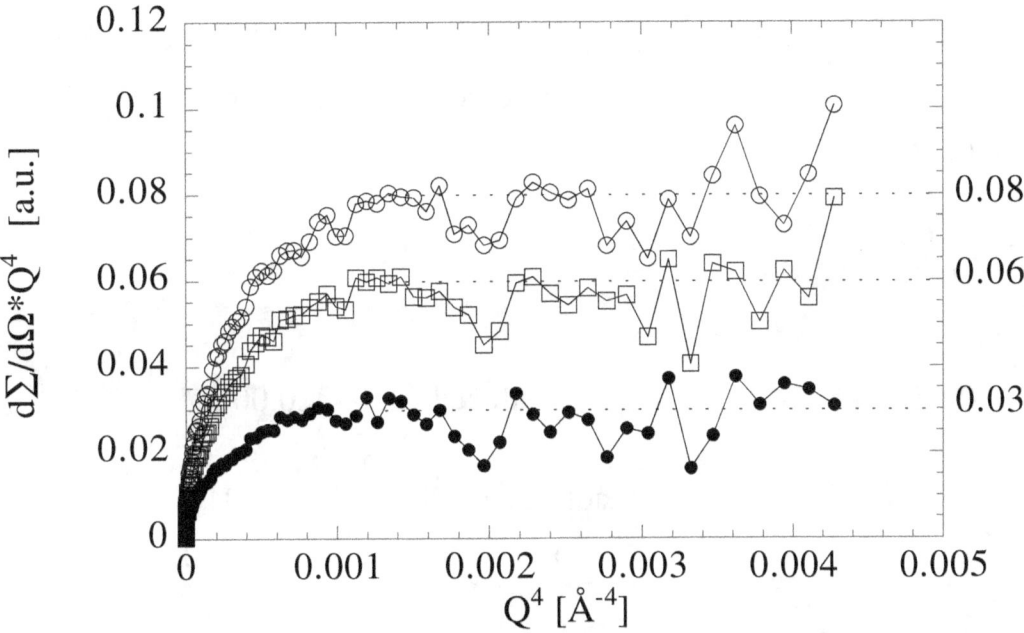

Figure 6.44: Porod plots (from SANS) of electrochemically treated SIGRADUR®G with 60 μm thickness. Filled circles denote the non activated samples. Open circles denote the oxidized samples. Open squares denote the oxidized and subsequently reduced samples. The Porod constants are written in arbitrary units. Oxidation time was 1 hour.

measure for the internal surface area of the samples. Upon oxidation, the internal surface area of the whole sample increases by a factor of around 2.7 from 0.03 to 0.08 [a.u.]. However, after reduction the internal surface area decreases from 0.08 to 0.06 [a.u].

6.4. COMPARISON WITH ELECTROCHEMICALLY ACTIVATED GLASSY CARBON

These values must be corrected, because the scattering data of the treated GC sheets are obtained from sandwichlike samples. However, the film thickness was not known. Therefore only an estimation for the contributions of core material and film material was possible.

As a rough estimation, the Porod constant of the untreated sample may be subtracted from the Porod constants of the treated samples. Therefore, the ratio of the Porod constants of the oxidized film (P_{ox}) and the reduced film (P_{red}) may be estimated as follows: $(0.08-0.03)/(0.06-0.03) > P_{ox}/P_{red} = 1.67$.

The ratio of internal surface areas prior to and after reduction (1.67) matches the ratio of the capacitance prior to and after reduction reasonably well (1.86).

Figure 6.47 displays a Porod-plot of SAXS curves concerning a couple of samples activated 30 minutes at 2.07 Volt and reduced at -0.5 Volt. The Porod-plot of

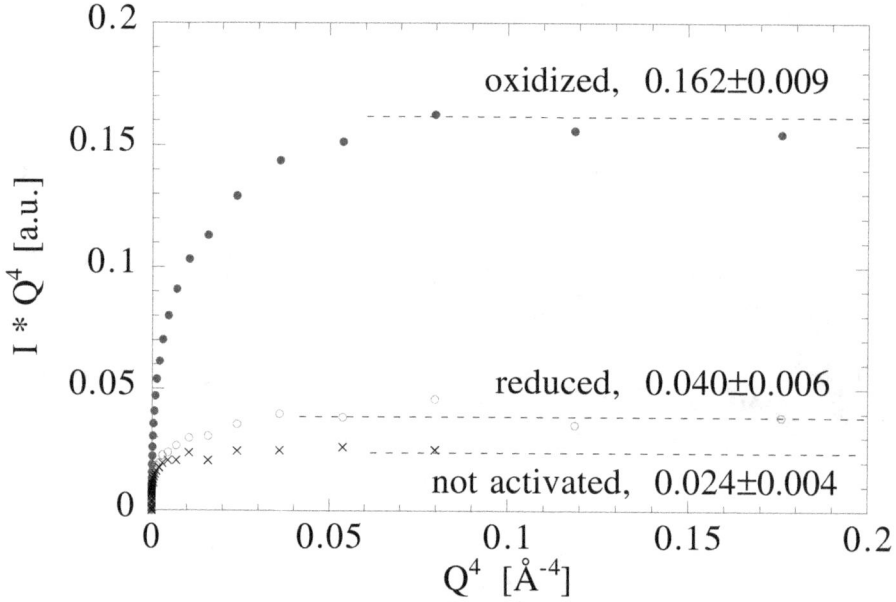

Figure 6.45: SAXS curves of electrochemically treated GC G with a thickness of 60 microns. Activation was 30 minutes at 2.07 Volt. Reduction was 15 minutes at -0.3 Volt.

the SAXS curves qualitatively confirms (i) that the internal surface area increases after oxidation and that (ii) the internal surface area decreases after subsequent reduction. The corrected ratio of Porod constants, however, is larger than in the case of the samples activated 1 hour and measured with SANS: $P_{ox}/P_{red} \approx 5 - 10$. The fact that the sample was activated only 30 minutes (SAXS measured) cannot explain the larger ratio ($\approx 5 - 10$) between the internal surfaces than in the case

Sample thickness and density (1h activated, 15 min reduced)						
	Thickness [μm]			Mass density [$mgcm^{-2}$]		
Pressure	non-act.	act.	act.+red.	non-act.	act.	act.+red.
without	55-56	106±13	77.5±12			
with	47-48	67±1	55±4	13.6±0.5	14±0.2	10.9±0.3

Table 6.13: Sample thickness and mass per sample area for non-activated and electrochemically treated thin SIGRADUR®G sheets with 60 microns nominal thickness. Activation was one hour at 2.07 Volt. Reduction was 15 minutes at -0.5 Volt.

of the sample activated 1 hour (SANS measured). It has to be taken into account that the neutrons in SANS are very sensitive to hydrogen, and maybe presence or absence of hydrogen causes differences in the two measurement techniques.

The internal surface area as measured with SAS concerns the geometrical internal surface area, regardless whether it is accessible to gases and liquids, or not. As the capacitance decreases to the same extent as the internal surface area, it may be concluded that the pores are not simply closed by some effect. More probably, a collapse of the pore network upon the reduction may be one explanation for the decrease of the internal surface area and capacitance.

During electrochemical treatment (oxidation and reduction), the sample thicknesses changed remarkably. Thicknesses were measured with a Stylus profile detector and are listed in Table 6.13. Before thickness measurements, the sample were dried in vacuum at 70°C. As the thickness depended on whether the detector tip on the sample was pressurized, the thickness was measured without a pressure and with a manually applied pressure, which could be quantified only qualitatively by the operator. After oxidation, the sample thickness increased by around 40% (100%, when no pressure was applied). After reduction the thickness decreased again to a value which was still higher than the original thickness.

As the sample experiences drastic volume changes during oxidation and reduction, an according degradation is expected.

The diffusion resistance also increased after reduction, as listed in Table 6.14. As the decreased capacitance after reduction was found in situ when samples were still in electrolyte, the decrease of the internal surface area is not necessarily caused by the drying of the samples. The decreasing capacitance and internal surface area, the volume and mass changes and the increasing diffusion resistance after reduction support the suggestion that a collapse of the pore network occurs during reduction.

6.4. COMPARISON WITH ELECTROCHEMICALLY ACTIVATED GLASSY CARBON

Diffusive resistance [$m\Omega cm^2$]		
Bias potential	act.	act. + red.
0.0 V	186.6	296.1
0.4 V	189.8	304.0
0.9 V	185.6	285.3

Table 6.14: Diffusion resistance of electrochemically treated SIGRADUR®G sheets with 60 microns nominal thickness for different bias potential. Activation was one hour at 2.07 Volt. Reduction was 15 minutes at -0.5 Volt.

Finally, the SAXS curves of the samples activated for 30 minutes were plotted in a Guinier plot (Figure 6.48). Unlike the SIGRADUR®G samples with 1 mm thickness (Figure 5.4, Chapter 5.2), the samples with 60 microns nominal thickness exhibited no Guinier curvature in the micropore region. The pore diameter of the non-oxidized sample was 24.37 Å. After oxidation the pore diameter decreased nearly by a factor of two to 12.40 Å. After subsequent reduction, the pore diameter increased again to 24.06 Å. The pore diameters mentioned should not be taken absolutely serious, because they were derived by a Guinier plot of scattering curves which exhibited not the typical features of a Guinier range.

However, it is obvious that the scattering curves of the non-activated sample and the sample which was activated and reduced subsequently, are nearly identical, while the activated sample remarkably differs from the two former curves.

In particular, the slope of the SAXS curve of the activated sample is smaller than the slope of the two other samples. This reveals that the pores in the activated sample are smaller than the pores in the two other samples.

Possibly oxidation causes a narrowing of pores - maybe by decorating the pore walls with surface functional groups -, and subsequent reduction removes these surface groups, causing a pore enlargement.

However it seems that even the pore enlargement after reduction does not compensate that the internal surface area and the capacitance decrease.

SIGRADUR®G sheets with 60 microns nominal thickness and the corresponding disks with 1 mm thickness deviate considerably from eachother with respect to their behaviour after electrochemical oxidation and reduction.

Possibly structural differences explain why the 1 mm samples keep their consistency during electrochemical activation and reduction, while the thin sheets experience a drastic degradation.

The structural differences are obvious from the XRD diffractograms: The thin sheets have larger and sharper peaks and therefore larger crystallites.

The SAXS and SANS curves of the 1 mm samples clearly show typical glassy

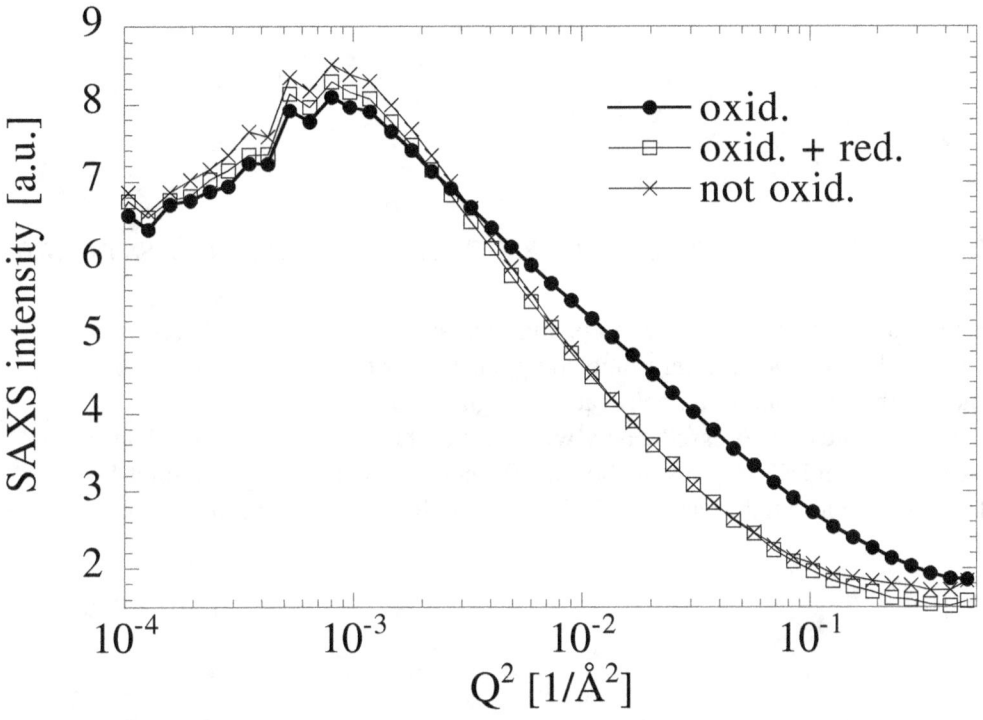

Figure 6.46: SAXS Guinier plot of electrochemically treated thin SIGRADUR®G sheets. State of sample is marked with symbols accordingly in the legend. Activation was 30 minutes at 2.07 Volt. Reduction was at -0.5 Volt.

6.4. COMPARISON WITH ELECTROCHEMICALLY ACTIVATED GLASSY CARBON

carbon structural features (Guinier range and plateau), while the thin sheets only show a power law, but no Guinier range and plateau. It is therefore very questionable whether the thin sheets can be regarded as glassy carbon.

Finally, the sheets can be pressed easily with a tip by hand so that their thickness changes from around 55 microns to below 50 microns. This material cannot be regarded anymore as a hard carbon. Glassy carbon, however, is definitely a hard carbon.

6.5 Structure and Structural Changes in GC

6.5.1 Structural Differences in GC

Figure 6.49 displays diffractograms of four types of GC with pyrolysis temperatures of 800°C, 1000°C, 1250°C and 2200°C. These samples have rather the same background intensity I_{back}, but the (002) peak and the (101) peak have a higher intensity, when the pyrolysis temperature was higher. The R-ratios (mentioned in

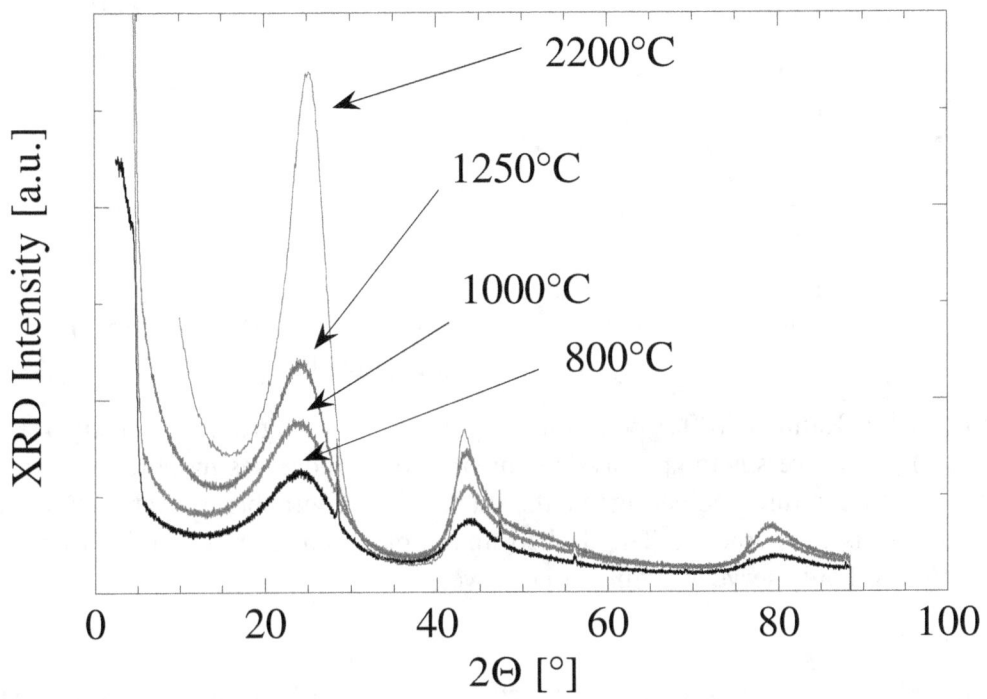

Figure 6.47: X-ray diffractograms of GC with 1 mm thickness and different pyrolysis temperature, as marked in the plot. The K800, K1000 and G2200 are GC samples as received from the manufacturer. K1250 was obtained by after-pyrolysis of the K800 samples. Samples with a higher pyrolysis temperature have a higher peak intensity. The (002) peak of the K-type GC samples is located at around 24.2°, yielding a lattice spacing of 3.68±0.1 μm. The (002) peak of the G2200 GC is found at 25.15°, yielding a lattice spacing of 3.54±0.1 μm.

Chapter 5.1) of all these samples are plotted versus the pyrolysis temperature in Figure 6.50. With increasing HTT the R-ratio increases as well. It was experimentally observed [30] that samples from hard carbons with a small R ratio have smaller nanopores than those with a large R ratio.

Obviously the stack height L_c of graphene sheets increases with pyrolysis temperature, because the R-ratios increase for samples with increasing pyrolysis temperature and therefore less single graphene sheets are present in the GC. The FWHM

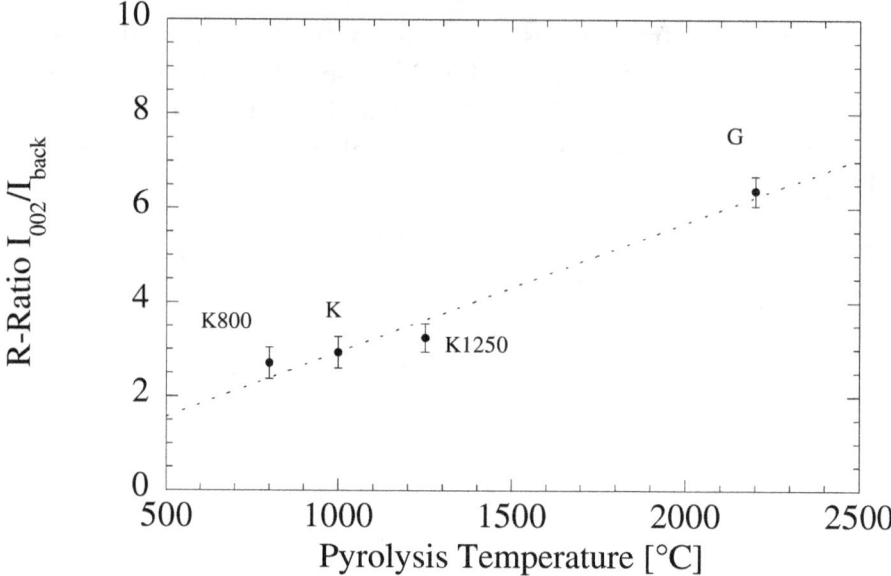

Figure 6.48: Ratio of (002) peak intensity I_{002} and background intensity I_{back} (R-ratio) plotted versus the pyrolysis temperature. R increases monotonous with pyrolysis temperature, indicating that the stack height of graphene sheets increases with pyrolysis temperature. The dashed line is obtained from a weighted linear fit, but should only serve as a guide to the eye.

of the (002) peak of the samples is plotted versus the HTT in Figure 5.48. As the FWHM is decreasing with HTT, an increasing mean value for L_c is indicated for increasing HTT, as expected [16]. The corresponding values for L_c are plotted in Figure 5.49 versus the HTT. While the samples with a HTT of 800°C have a value for L_c of around 9 Å, the value for the samples with a HTT of 2200°C is nearly doubled. The data from Short and Walker [58] are between 13 Å and 24 Å (see Table 5.1) and are therefore in quantitatively good agreement with the data found in this thesis. Care must be taken in interpreting the data because the true values (as obtained from TEM [16]) for L_c are probably by a factor of 10 higher. However, the expected trend of enlarging stack height takes place upon heat treatment.

The Scherrer formula can also be applied to the (100) reflex or (101) reflex of GC, which are found around $2\Theta \approx 44°$. With these values the lateral crystallite size L_a can be determined. For the 1 mm K-type GC samples a value of around 20 Å was found, while the corresponding G-type samples had a lateral crystallite size of around 27 Å.

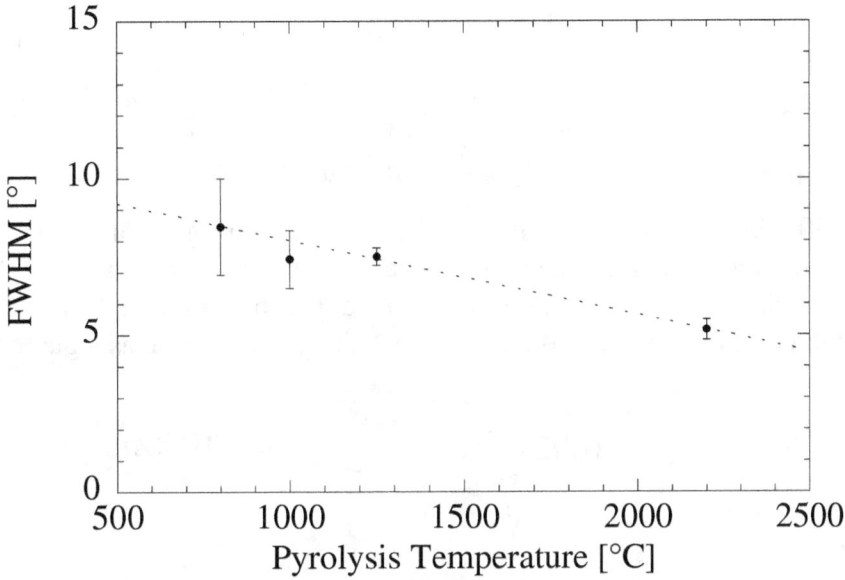

Figure 6.49: Full-Width-at-Half-Maximum (FWHM) of the (002) peak in GC as a function of pyrolysis temperature. The FWHM decreases with increasing pyrolysis temperature. The dashed line is obtained from a weighted linear fit, but should only serve as a guide to the eye.

6.5. STRUCTURE AND STRUCTURAL CHANGES IN GC

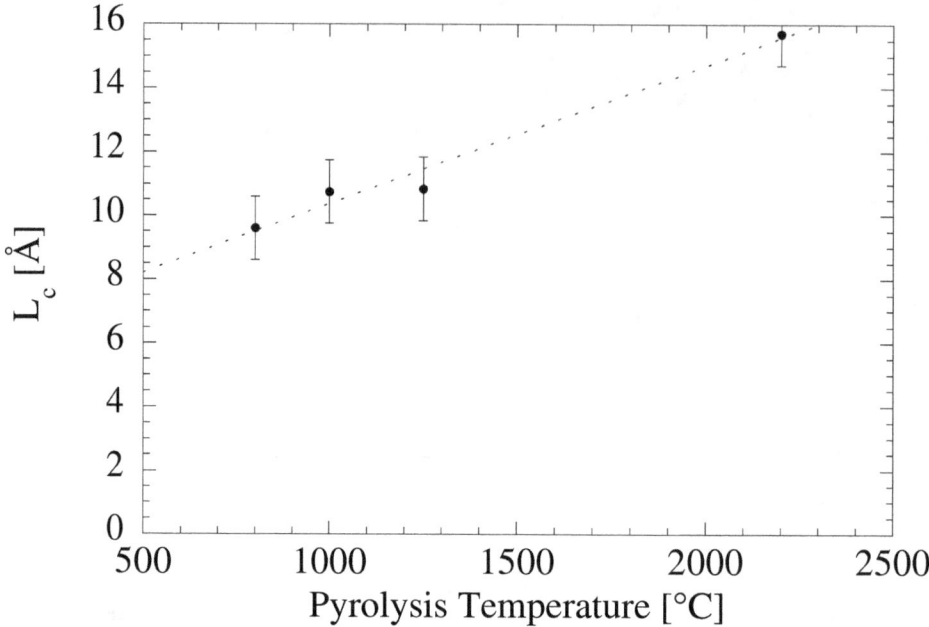

Figure 6.50: The vertical crystallite size L_c increases during pyrolysis, when the temperature increases. In the temperature range between 800°C and 2200°C, L_c increases by around a factor of two from around 9 Å to around 16 Å. The dashed line is obtained from a weighted linear fit, but should only serve as a guide to the eye.

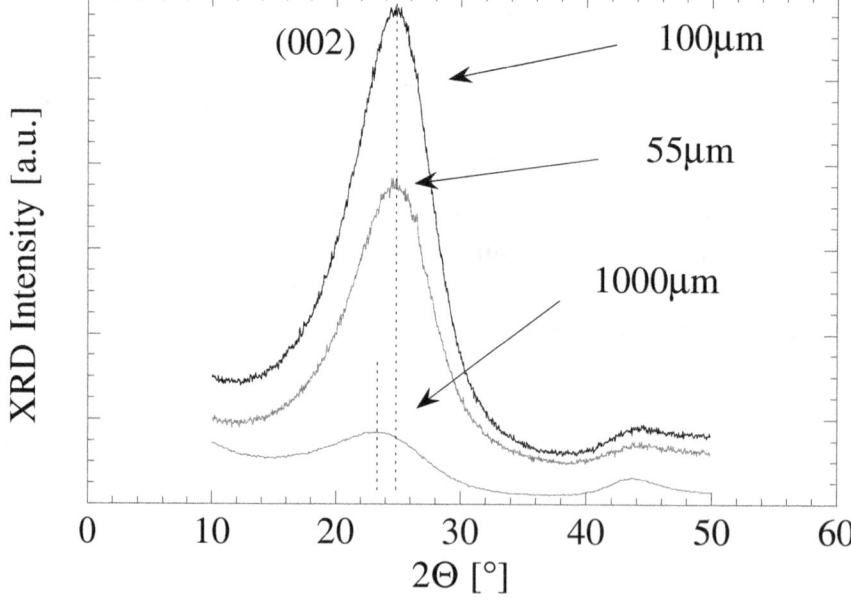

Figure 6.51: X-ray diffractograms of GC with 1000 μm (lower curve), 100 μm (upper curve) and 55 μm thickness (middle). Only the (002) peak is well developed. The 55 and 100 μm samples have the peak maximum at a larger angle (24.67°) than the 1000 μm sample (24.67°). For clarity, dashed curves denote the positions of the maxima.

Finally it is remarked again that the position of the (002) peak is not necessarily at a fixed position. Moreover, the peak position shifts to larger diffraction angles with increasing HTT, indicating that the graphene layers become closer with increasing HTT.
The shift is particularly pronounced when comparing K-type and G-type GC.
With these experimental findings it has become clear that important part of the structure of GC is determined by the pyrolysis temperature.

Structural differences were also found for samples with the same HTT, but different sample thickness.
Originally there should be no dependency of the structure on the sample thickness. However, the pyrolysis schedule of the manufacturer was not available, and it is possible that the samples were pyrolyzed for different long times.
To create GC of a certain type (K, G), a minimum duration time for the pyrolysis is necessary. This minimum depends exponentially on the sample thickness [121]. This is the reason why very thick GC samples (more than 1 cm) are not at all available on the market, because it would take years of HTT. In contrast, very thin GC sheets of some 10 microns are prepared within some hours.
For many applications only the surface properties of GC are important, for instance in applications where corrosion occurs. Therefore it is not very important when surface structure differs from bulk structure.
There is evidence from XRD and SAXS that the 55 micron samples have a different structure than the 1000 micron samples. Figure 6.53 displays XRD diffractograms of untreated K-type GC samples of 55, 100 and 1000 microns thickness. While the 55 and 100 micron samples have the maximum of the (002) peak at $24.67°$, the 1000 micron sample has the peak maximum at around $23.33°$, which indicates that, according to Bragg's law, the graphene sheets are closer in the thin GC samples. The corresponding interlayer distances d are 3.61 ± 0.1 Å for the thin samples and 3.81 ± 0.1 Å for the $1000 \mu m$ sample.

The full width at half maximum of the XRD diffractograms of the thin samples is also somewhat smaller than that of the 1000 micron samples, which indicates, after the Scherrer formula, that the graphene sheets of the thin samples are laterally more extended than those of the 1000 micron samples.
To complete the picture, a diffractogram of GC pyrolyzed at 2200°C with 46 microns thickness (G-type GC sheets with 60 microns nominal thickness) is displayed in Figure 6.54.
The XRD intensity in counts/second is plotted on a logarithmic scale, because the (002) peak, whose position is found at $26.33°$ with a lattice distance of $3.38 \mu m$, dominates the other peaks by far.
Evaluation of the FWHM allowed to estimate the crystallite size to $L_a \approx 430$ Å and $L_c \approx 540$ Å. Without taking these values too serious, they nevertheless serve

6.5. STRUCTURE AND STRUCTURAL CHANGES IN GC

XRD data for various GC			
Type GC	2Θ °	d_{002} [Å]	ρ_x
G 60 μm	26.4	3.3732	2.2426
G 1mm	25.3	3.5173	2.1507
G 3mm	25.5	3.4902	2.1674
K 100 μm	24.9	3.5729	2.1172
K 60 μm	24.9	3.5729	2.1172
K800 110 μm	24.5	3.6304	2.0837
K 800 1mm	23.8	3.7355	2.0251

Table 6.15: Position of the (002) peak, interlayer distance and X-ray density of various kind of GC with different HTT and thickness, as obtained from XRD.

as a rough qualitative estimation for the extension of the crystallites. Probably this material cannot be considered anymore as glassy carbon. Rather it should be considered as turbostratic graphite with a pronounced lamellar structure. In the section about electrochemical activation (Chapter 6.4), GC of this type was presented and discussed. The G-type GC samples show the most pronounced difference in structure concerning their different thicknesses. Figure 6.55 summarizes the results found for the lattice spacing of various types of GC studied in this thesis. The values are compared with data found by Short and Walker, showing a steep decrease of lattice spacing d with increasing HTT in the range from 500 to 1000°C and a less pronounced, but nonetheless remarkable decrease for higher HTT.

The interlayer distance of graphite (3.33538 Å) is placed arbitrarily at 3000°C as a reference.

The change of lattice spacing of GC samples studied in this thesis is in qualitative agreement with the values found by Short and Walker. In particular, the decrease of interlayer spacing with higher HTT is verified.

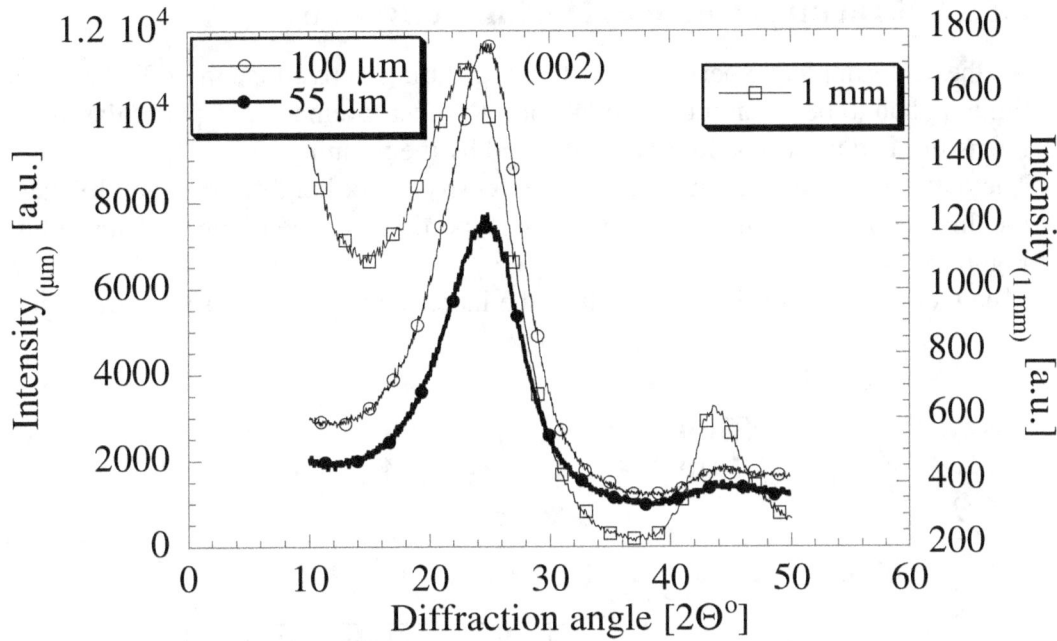

Figure 6.52: X-ray diffractogram of GC pyrolyzed at 2200°C and 46 microns thickness. Intensity is plotted on a logarithmic scale because of the dominant (002) peak, whose position is found at 26.33° with a lattice distance of 3.38 Å.

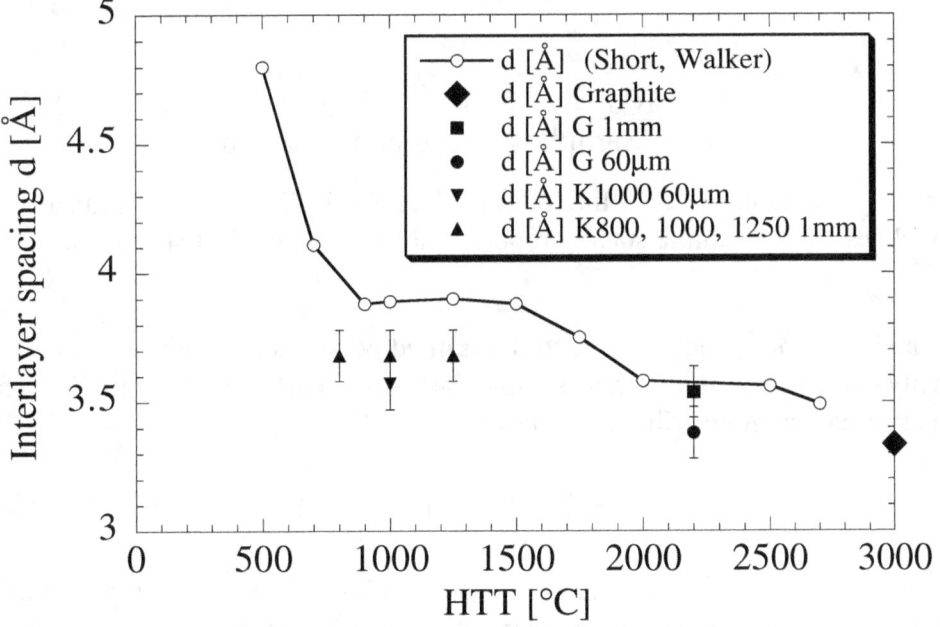

Figure 6.53: d-spacings for GC as a function of the HTT. Data from literature ([58, 16]) are compared with own data. With increasing HTT, the d-spacing decreases monotonously, as verified experimentally.

6.5.2 Structural Changes During Activation

As the GC samples experienced a thinning during activation, the effect of this thinning had to be taken into account for XRD measurements. In particular, peak intensities and peak positions were affected by the thinning.

The data for the maximum (002)-peak intensity (peak height) are plotted versus the thickness of the sample stack (this is proportional to the number of sheets) in Figure 6.56.

The increase of intensity is a result of the increasing number of scatterers, when

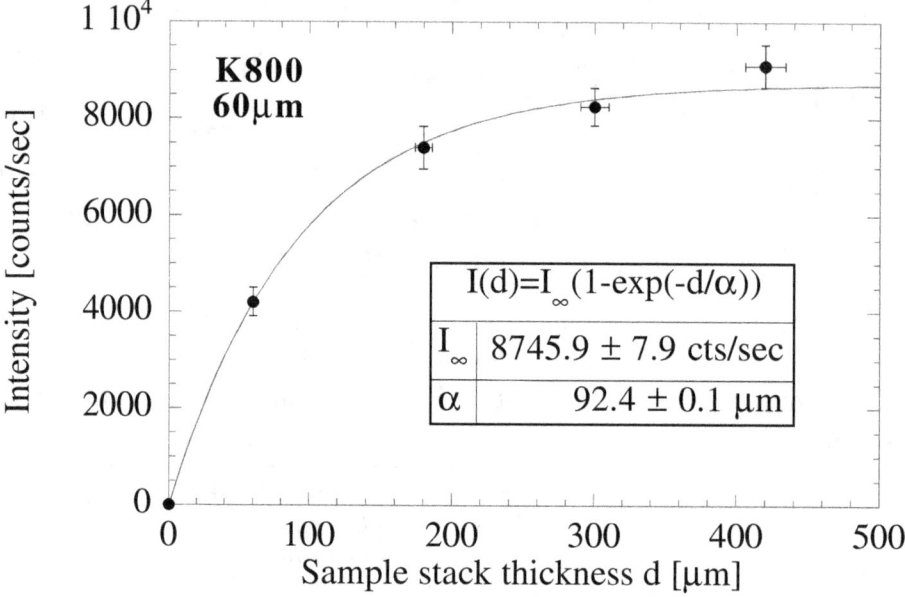

Figure 6.54: XRD intensity of the (002) peak of K 800 GC sheets as a function of the thickness of the sample stack, proportional to the number of stacked sheets.

the samples become thicker. Due to the limited penetration depth of X-rays, the intensity remains constant when samples are sufficiently thick. The diffracted intensity I can be fitted with an expression of the form

$$I(d) = I_\infty \left(1 - exp\left(-\frac{d}{\alpha}\right)\right) \tag{6.62}$$

d being a measure for the thickness, α a penetration depth and I_∞ the diffracted intensity, when the sample can be regarded as a semi-infinite slab.

The diffracted intensity increases with the number of scattering atoms, but also decreases as a consequence of absorption. As the samples investigated with XRD are mainly thin (30 to 60 μm thickness), it was assumed that the influence of this effect on the measurements could be estimated linearly. But this effect was negligibly small. The (002) peak position is also depending on the sample thickness,

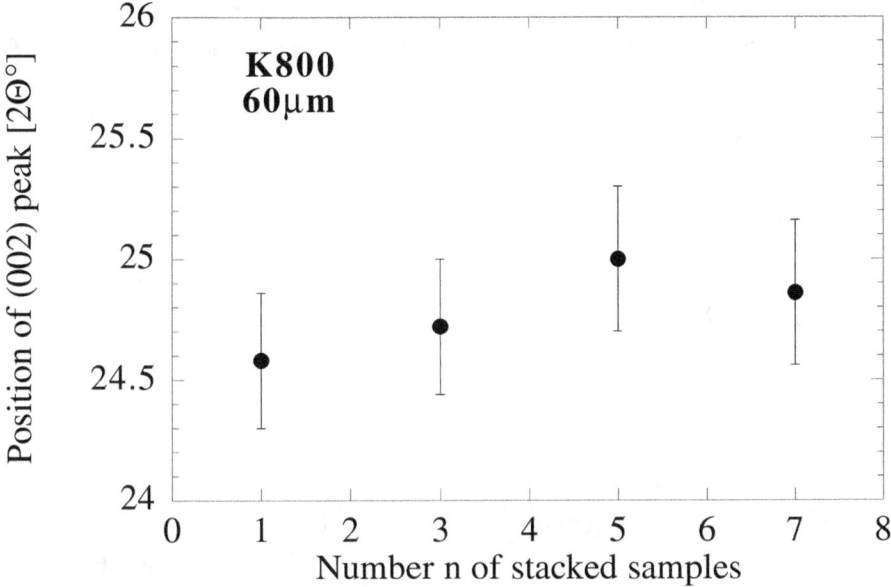

Figure 6.55: Position of the XRD (002)-peak as a function of the sample stack height. The number of sample sheets from K 800 with 60 microns thickness is denoted n in the ordinate.

which can be extracted from Figure 6.57.

For growing sample thickness the (002) peak shifts toward larger diffraction angles. Therefore it can be excluded that the peak shift after oxidation and sample thinning towards larger diffraction angles is caused by a change of the sample thickness.

When the samples become thicker, the peak is shifted to larger values of 2Θ. This effect is visible only over a range of several hundred microns.

During activation, however, the samples become thinner (in 3 hours only around 30 microns), and the (002) peak shifts nevertheless to larger 2Θ values. This is a direct proof that the shift of the (002) peak towards larger 2Θ values is not caused by an artifact as mentioned before (change of sample thickness or amount of material, respectively).

It is a particular problem in evaluating the XRD data that the samples consist of film material and bulk material, and thus the diffracted intensity includes contributions from the bulk material which we so far were unable to separate. Therefore results on samples either not-activated or fully activated are presented here, although also *sandwiches* were measured.

6.5. STRUCTURE AND STRUCTURAL CHANGES IN GC

It could also be possible that the unactivated sample has already some structural differences or anisotropy along the cross section [110].

In Figure 6.58, the (002) peaks of a non-activated GC sample and of a GC sample activated for three hours are displayed. The (002) peak of activated GC is found

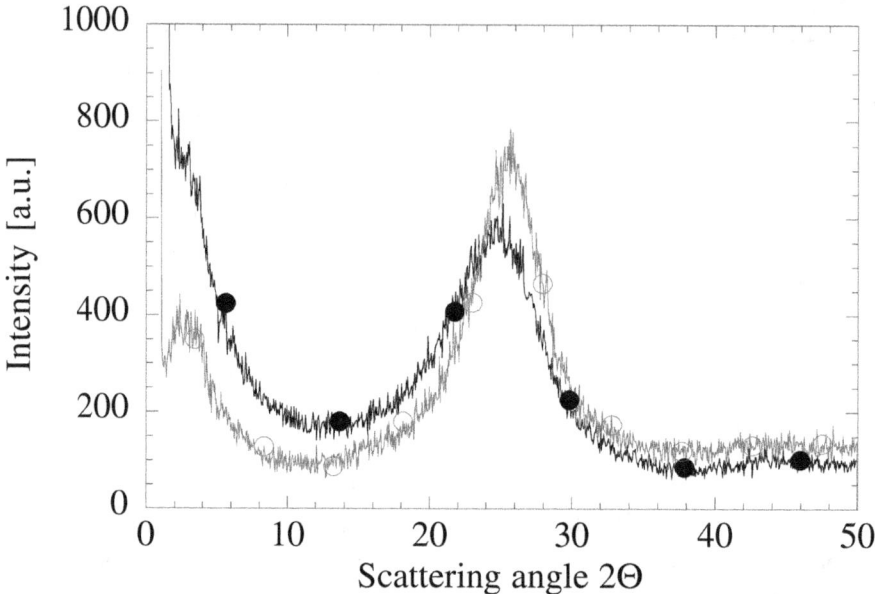

Figure 6.56: X-ray diffractogram of a non activated GC sample (•) and a GC sample activated for 3 hours (o). The (002) peak around 25° is shifted towards larger 2Θ angles after activation and has a smaller full width at half-maximum intensity (FWHM).

at a larger diffraction angle (25.7±0.5°) than the peak of the non-activated sample (24.9±0.5°). The shift indicates that the mean interlayer spacing between the graphene sheets decreases [122] upon activation (non-activated: $d = 3.57\pm0.07$ Å; activated for 3h: $d = 3.46\pm0.07$ Å). Therefore, the X-ray density or skeleton density ρ_x of the GC also increases upon activation according to [48]

$$\rho_x = (3.33538/d_{002}) \times 2.268 g/cm^3, \tag{6.63}$$

where 2.268 g/cm^3 is the density of graphite with d_{002}=3.33538 Å. The peak intensity in the diffractogram of activated GC is also higher than for non-activated GC, and the full width at half maximum (FWHM) value B for the (002) peak is decreasing upon activation (7.1±0.5° non activated; 5.4±0.5° 3h activated). Obviously the remaining GC has a denser structure, probably by removing less ordered carbon atoms in the GC due to burn off. The increase of structural order of GC powder upon activation was already mentioned elsewhere [123]. Applying the Scherrer formula for the crystallite size L_c [57], an increase of the remaining average graphene layer stack height L_c from about 11.5±0.8 Å to 15.4±1.4 Å is found. From the relationship between crystallite size and layer distance,

$$L_c = d(n-1) \tag{6.64}$$

we conclude that the mean number n of stacked graphene sheets grows from 4.2±0.2 to 5.4±0.5, which behavior is in line with the *falling cards model* suggested recently by W. Xing et. al. [30].
This model describes how an arrangement of graphene stacks in hard carbons could be affected by a thermal oxidation process. Glassy carbon is built up from graphene layers, which enclose tiny pores of a few nanometer in size. The graphene sheets are arranged like polyhedra, and the empty space inbetween are the pores [25].

By oxidation interlinking atoms between the graphene layers are burnt off, and the graphene layers have enough kinetic energy to rotate or to rearrange parallel to another stack of graphene layers.
As a result, if the rotating layers were separating two pores before, now the separation is lifted and a pore larger than the pore before is created.
In this thesis the porosity p of the non-activated GC was determined from its X-ray density ρ_x (Chapter 5.1) and its apparent density ρ_g, which can be obtained from the weight and the dimensions of the sample, by using the relation [48]:

$$p = 1 - \frac{\rho_g}{\rho_x} \tag{6.65}$$

and found a value of $p = 23.0 \pm 1.5\%$ with $\rho_g = 1.63$ g/cm^3 and $\rho_x = 2.11$ g/cm^3. For GC activated for 3h, with an apparent density of $\rho_g = 1.21$ g/cm^3 and $\rho_x = 2.18$ g/cm^3, a porosity of $p = 44.6 \pm 4.6\%$ was found.
A change in the X-ray density ρ_x will influence the scattering contrast Δn_f of the GC, which is of importance for the evaluation of SAXS data when absolute intensity data are measured [67], because the scattered SAXS intensity is proportional to Δn_f^2. In the present work an increase of 4.5% was found in the scattering contrast Δn_f during activation, which results in a 10% reduction in the internal surface area for the sample activated for three hours. The determination of the scattering contrast is described in the experimental part to this work.
For samples activated not longer than 40 minutes, we observed that the (002) peak shifted to smaller diffraction angles (around 24.5°), suggesting some widening of the distance between graphene sheets. It is not clear yet which physical process causes this effect. Possibly the widening occurs exactly at the interface between

6.5. STRUCTURE AND STRUCTURAL CHANGES IN GC

film material and bulk material when oxidation takes place and graphene sheets begin to reorientate.

6.6 Small Angle X-ray Scattering on Glassy Carbon

6.6.1 SAXS on Various Non-Activated GC

Figure 6.59 displays the SAXS scattering curves of two not activated GC samples with 1 mm thickness (SIGRADUR®K and SIGRADUR®G) in a Guinier-plot [65]. The plateau region is clearly pronounced for both samples, and the plateau height yields rather straight the scattering at zero angle for the micropores.

The plateau of the SIGRADUR®G sample has a higher intensity than that of the SIGRADUR®K sample.

The scattering at zero angle, $\Delta n_f^2 N V^2$, varies with the sixth power of the sphere radius.

The scattering at zero angle is therefore more sensitive to a change in pore size than to a change in the pore number [30].

The slope of the scattered intensity for large scattering vectors Q is different for

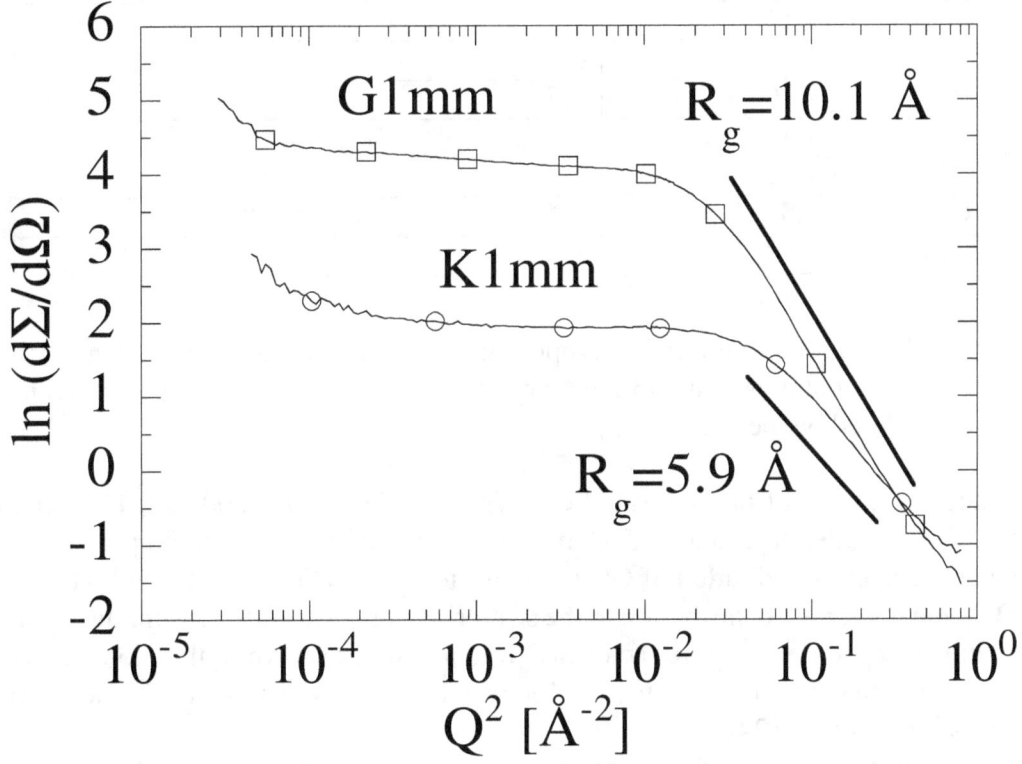

Figure 6.57: Guinier-plot of non-activated (SIGRADUR®K and SIGRADUR®G, 1 mm thickness). The radius of gyration R_g was determined from the slope of the curve for $Q^2 \geq 8 \times 10^{-2}$.

6.6. SMALL ANGLE X-RAY SCATTERING ON GLASSY CARBON

both samples. The slope of the G sample is larger than the slope of the K sample. Considering the Guinier approximation for the SAXS curves, it becomes qualitatively clear that the G sample has larger pores than the K sample.

For a quantitative analysis, a Guinier tangent was applied to the curve with the result that the radius of gyration is R_g=5.9 Å for the K sample and 10.1 Å for the G sample.

As a first quantitative result we find that the micropores of G samples are nearly as double as large (radius R=7.8 Å) as the micropores of the K samples (radius R=4.6 Å).

This result is qualitatively expected because GC with a higher HTT has larger micropores [16].

Table 6.16 lists the micropore gyration radius and the mass density of GC depending on the heat treatment temperature.

H.T.T [°C]	Density [g/cm^3]	R_g [Å] (A.C. Craievich)	R_g [Å] (PSI [65])
500	1.14	-	-
600	1.17	4.2	-
700	1.26	4.7	-
800	1.43	7.0	-
900	1.52	7.9	-
1000	1.55	8.4	5.9
2200	-	-	10.1

Table 6.16: Mass density and micropore gyration radii of GC, depending on the HTT. Data in columns 2 and 3 are taken from literature [16]. Radii in column 4 were measured by the author [65].

Scattering curves of non-activated GC types with HTT of 2200°C, 1000°C and 800°C were fitted to equation 5.16 in order to get information on the pore radius and pore radius distribution of GC with different HTT (Figures 6.60 and 6.61). Only the micropore contribution to the scattering curve was fitted assuming spheres, as indicated by the structure factor S(Q,R) for spheres shown in the plot. The radius in the maximum of the distribution was R_0 = 4.5 Å. The width of the distribution was σ=0.4 [124].

Logarithmically normal distributed spherelike voids were assumed. The scattering at zero angle, the radius R_0 in the maximum of the distribution and the width σ of the distribution were varied.

Pore size distributions are plotted in Figure 6.61, together with results found for activated and non-activated GC sheets of 60 microns thickness.

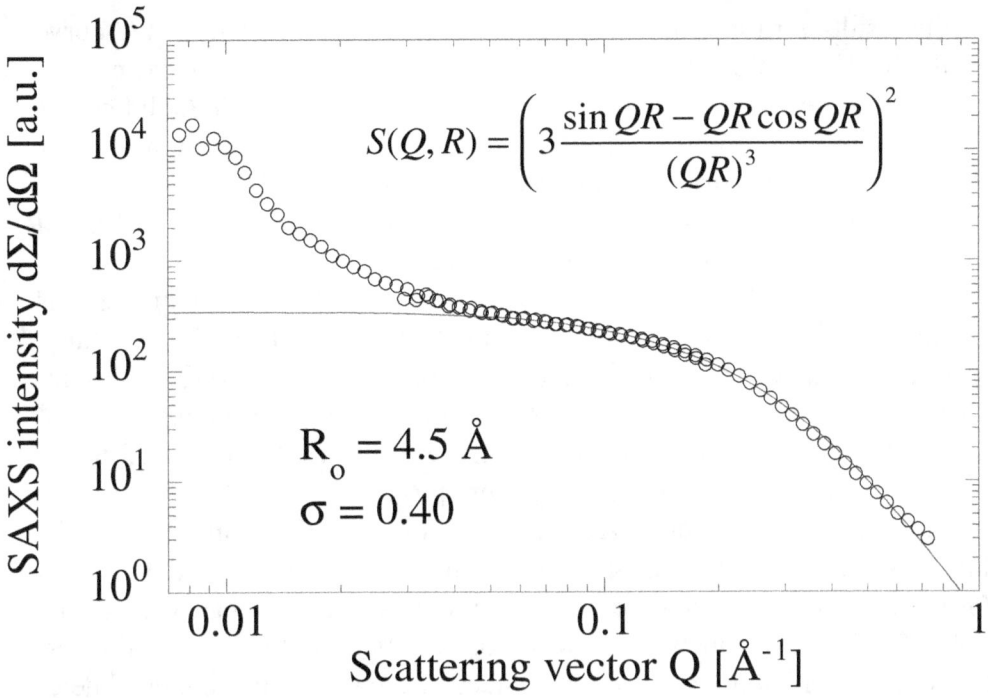

Figure 6.58: SAXS data and fitted scattering curve for micropores of a 60 micron thick samples (3 hours activated at 450°C).

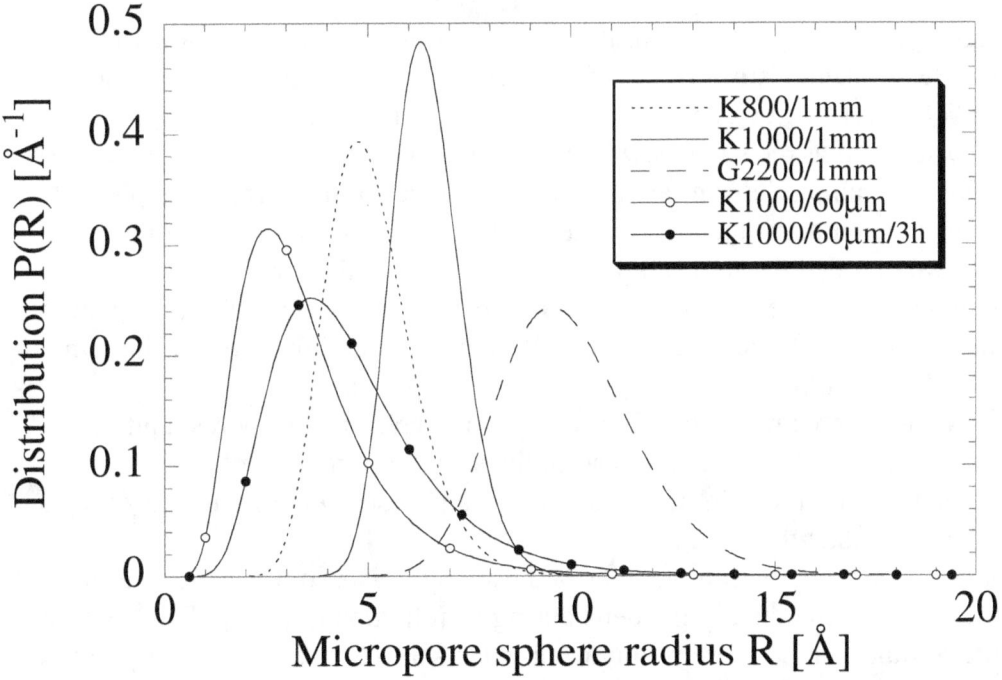

Figure 6.59: Lognormal distribution of pore sizes of various GC samples, which had received different HTT. Curves with symbols concern SIGRADUR®K with 60 microns thickness. The curve with the closed symbols concerns an activated sample (3 hours at 450°C).

6.6. SMALL ANGLE X-RAY SCATTERING ON GLASSY CARBON

Only the results for the 1 mm samples are discussed here. The dotted curve concerns the K800 sample, which has the smallest pores. The peak maximum of the distribution curve is around 4.8 Å. The SIGRADUR®K sample (solid curve) has the peak maximum around 7.2 Å. Finally, the GC with the highest HTT (SIGRADUR®G, dashed curve) has the peak maximum around 9.5 Å.

The trend as obvious from the Guinier plot of SIGRADUR®K and SIGRADUR®G is also followed by the pore size distribution curves.

Comparing the distribution plots of the samples with 1 mm thickness, it is obvious, that the K 800 samples have the smaller value for R_0 (dotted line), SIGRADUR®K samples have larger R_0 (closed line), and SIGRADUR®G samples (dashed line) have the largest R_0. The radius of the micropores growths, when the HTT is raised. This result is in line with the results presented in the XRD part of this thesis and reported in literature [16].

Figure 6.62 displays the pore size distribution of K800 GC samples with different thicknesses. The thinnest sample with 62 microns thickness has the smallest value for R_0 and for σ. The sample with 110 microns thickness has a somewhat larger value for R_0 and a much larger width of distribution. Finally, the thickest sample has values which are between the values of the two thin samples. Therefore no systematic trend between sample thickness and pore size distribution is obvious.

A similar inconsistency is given in the XRD data, for example as shown for the SIGRADUR®K samples in Figure 6.53. In there, the 1 mm thick sample has the largest FWHM for the (002) peak and the smallest intensity.
The 100 μ sample has the smallest width for (002) and the largest intensity. Values for the thinnest sample (55μm) are between the values of the former samples.
Comparing the non-activated GC samples with different thicknesses (1 mm vs. 60 microns), we find that the thin sample has smaller pores than the thick sample.
We recall again that the width of the (002)-peak in the XRD diffraction pattern was smaller for the 60 micron SIGRADUR®K sample and broader for the 1 mm SIGRADUR®K sample.
Obviously samples with a larger L_c (60 microns) have smaller pores, and viceversa - the 1 mm sample has smaller crystallite size, but larger pores.
It was assumed that the thin samples had remained shorter time in the furnace during pyrolysis than thick samples.
Possibly gases leaving the sample interior during pyrolysis could disturb the structural ordering, which actually happens during heat treatment [125]. This disturbing effect should maybe more pronounced when samples are thick and pyrolysis time was not sufficiently long.

Finally micropore size distributions of GC having received different HTT or having different thickness are compared in Figure 6.63. The thin (62 microns) K800 GC sample has the smallest micropores.

The pore size distribution of the 1 mm thick SIGRADUR®K sample has a quite similar shape as the one of K800 with 62 microns. It only seems to be shifted by around 2 Å towards larger radii. The width σ of both samples is nearly the same. The thin version of the SIGRADUR®K (60 microns) deviates considerably from the thin K800 GC and from the 1 mm SIGRADUR®K, because the maximum of the distribution is found at larger radii and the width of the distribution is also much broader.

Additionally the distribution of a sample having a different precursor is displayed. As a precursor a commercial Capton foil was pyrolyzed (as by information from the manufacturer, HTW GmbH.). This sample had a thickness of 150 microns.

6.6. SMALL ANGLE X-RAY SCATTERING ON GLASSY CARBON

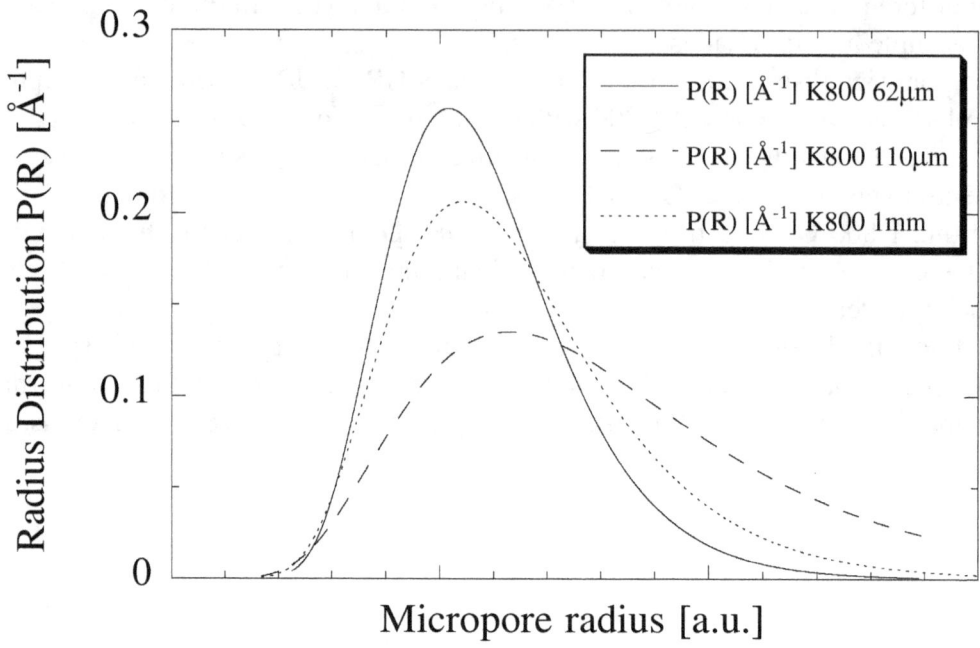

Figure 6.60: Micropore radius distribution of K800 GC with different thicknesses. Radii are given in arbitrary units. The samples have received the same HTT, but have different thickness.

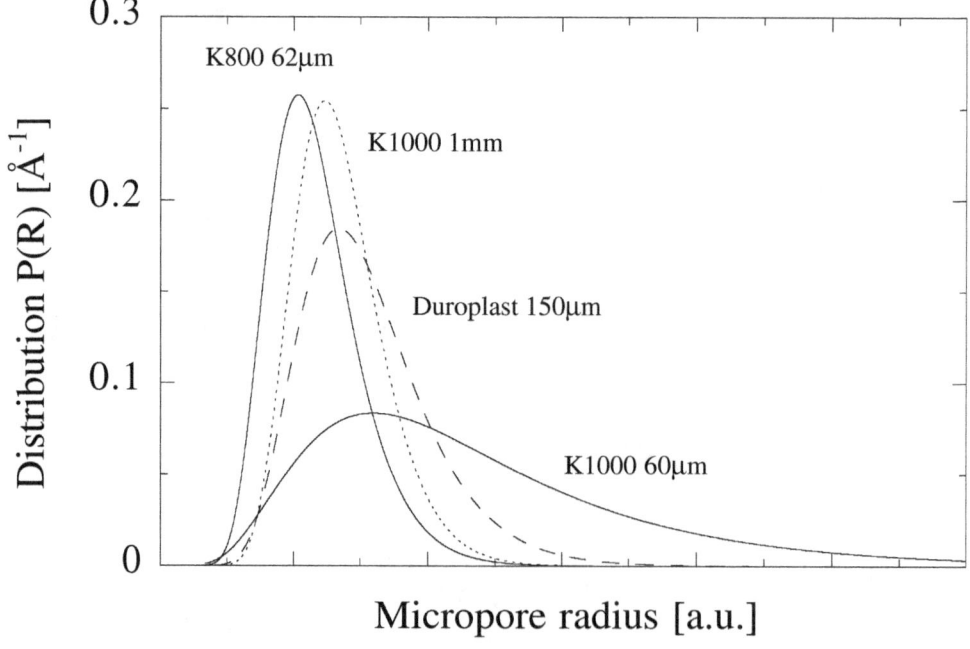

Figure 6.61: Radii are given in arbitrary units. None of the samples were activated.

6.6.2 SAXS on Thermally Activated GC

In order to study the influence of activation on the pore structure of GC, scattering curves from oxidized samples were recorded. It is emphasized that the scattering curves from the activated samples carry information both from unactivated GC core material and from the activated film material.

Therefore it is impossible to extract quantitative information from the raw data (scattering curves with no bulk material subtraction). But qualitative information can be inferred from the non-corrected scattering curves.

Figure 6.64 shows a log-log plot of scattering curves of SIGRADUR®K GC with 1 mm thickness [19].

Scattering curves of one non activated sample and two activated samples are displayed. The profile of the non-activated SIGRADUR®K is already known from Figure 6.59 and has a rather pronounced plateau region.

The scattering curves of the activated samples show a deviation of the plateau towards smaller Q values. The sample activated for 2 hours at 400°C has a slightly increasing intensity at the plateau region towards smaller Q values.

The film thickness of this sample is not known. But the film thickness can be estimated as follows:

Considering Figure 6.17 (capacitance of activated GC samples, SIGRADUR®K, 1 mm thickness) we find that the capacitance has nearly reached the region of saturation after 2 hours of activation at 400°C. The capacitance is between 250 and 300 mF/cm^2.

The capacitance of samples activated for 2 hours at 500°C is between 200 and 250 mF/cm^2, and the capacitance of samples activated at 450°C is between 400 and 450 mF/cm^2. Film thicknesses of the samples activated at 450°C and 500°C (Figure 6.12) are 60 and 30 μm.

From above results a film thickness of around 40 μm can be estimated for the sample activated 2 hours at 400°C. The burn-off rate was determined to be around 3 μm/h (see Chapter 6.1). The ratio of overall film thickness to unreacted core thickness is therefore less than 0.09 (¡9%).

The contribution of the film is less than 10%, but it has a significant influence on the scattering curve. This effect is even more pronounced for the sample activated at 500°C. The burn-off rate is 50 μm/h. So the sample has a thickness of 900 microns after 2 hours of activation. The film thickness is 30 microns on each sample side, so the ratio of film to unreacted core is around 7%.

---

[19]The samples were measured by Dipl.-Phys. Rainer Saliger and Dipl.-Phys. Reino Petricevic, Universität Würzburg

6.6. SMALL ANGLE X-RAY SCATTERING ON GLASSY CARBON

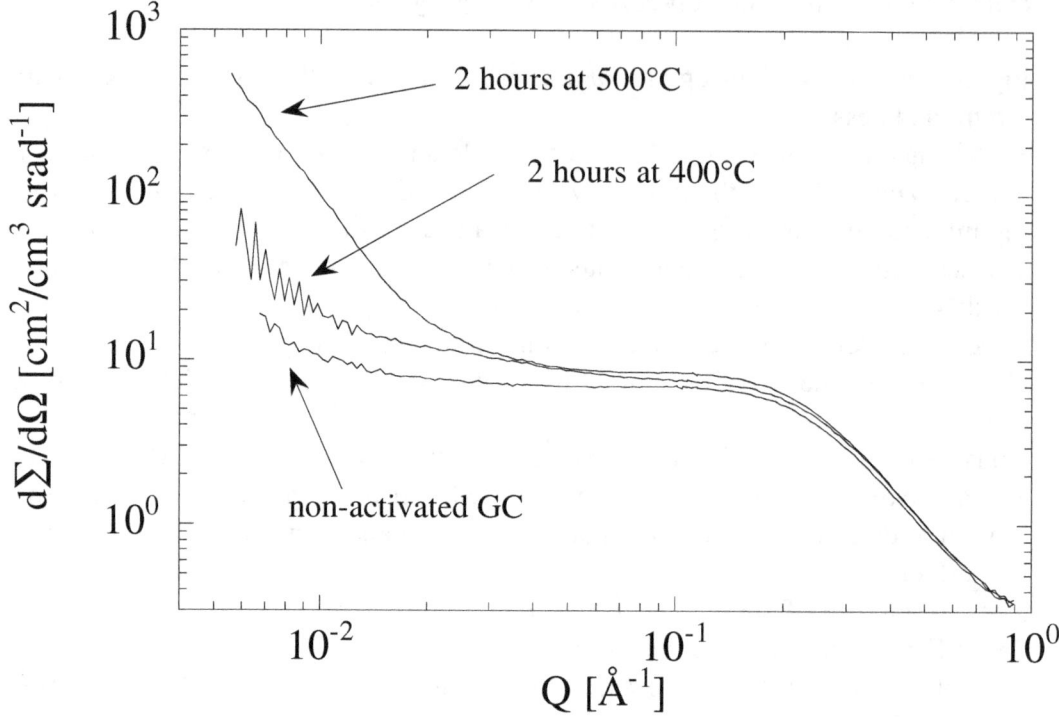

Figure 6.62: Scattering curves of non-activated and activated SIGRADUR®K samples with 1 mm thickness. The curves are raw data which had not received further treatment such as bulk- and background subtraction. Therefore the scattering curves exhibit both bulk- and film-properties.

As a qualitative result we find that the activation leads to higher SAXS intensity for small Q values. As Q is a measure for the reciprocal space, we may conclude that objects on a length scale from 10 Å to more than 100 Å are created by the activation.

Apart from these findings a very small change of the slope towards somewhat larger exponents of decay of the sharp decrease of intensity for large Q values is observed.

Probably the evolution of a network with open pores is monitored by the SAXS curves of the activated samples.

Figure 6.65 displays scattering curves of 1 mm SIGRADUR®K samples activated for different times at 450°C. The curves concern raw data which had not received

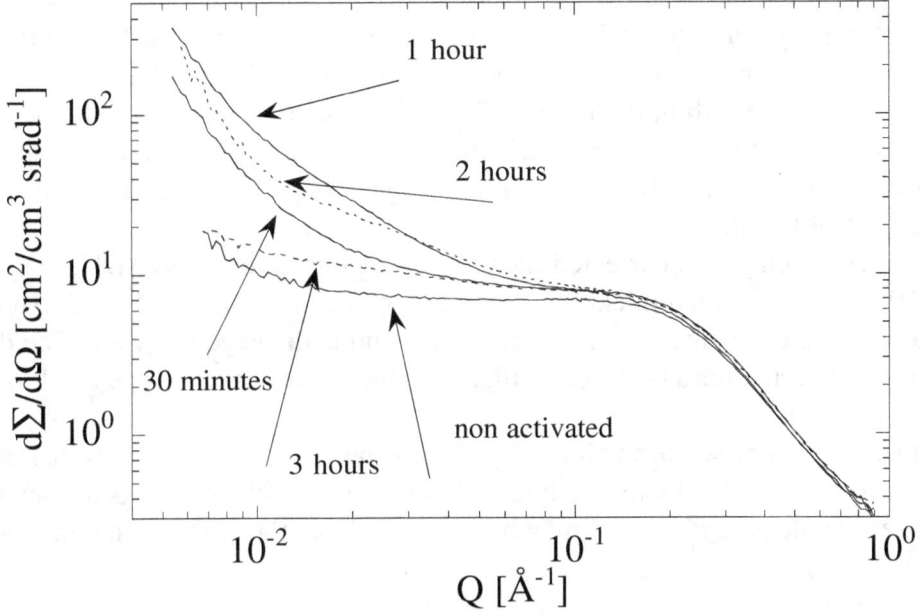

Figure 6.63: Scattering curves (raw data) of non-activated and activated SIGRADUR®K samples with 1 mm thickness. Activation temperature was 450°C. Activation times are denoted accordingly in the plot.

further treatment such as bulk- and background subtraction. Therefore the scattering curves exhibit both bulk- and film-properties.

There is a trend that the intensity for small Q values increases with activation time (after 30 minutes and after one hour of activation). After 2 hours of activation, however, a small decrease of the intensity is observed compared to the sample activated for 1 hour.

After 3 hours of activation, the scattered intensity at small Q values is nearly sim-

6.6. SMALL ANGLE X-RAY SCATTERING ON GLASSY CARBON

ilar to that of the unactivated sample. This decrease of intensity of this sample is directly correlated with a sharp decrease of the capacitance after 3 hours of activation (see Figure 6.17).

Figure 6.17 displays the capacitance of 1 mm SIGRADUR®K, activated at 450°C. After 3 hours activation, a sharp decrease in the capacitance is found. Probably large part of the film peels off after activation, which is also confirmed by the surface morphology diagram in Figure 6.8.

Although the ratio between film and bulk is less than 10%, changes are so strong that they are significant in the scattering curves. No quantitative analysis of the scattering curves of the 1 mm thick samples was possible, but it may be concluded that activation changes the pore structure, and objects are built relevant on a length scale starting from 10 Å towards larger dimensions.

To quantitatively study the influence of activation on the pore structure, thin GC sheets of 60 microns nominal thickness were prepared and characterized.

As the samples were thin, the ratio of film thickness to sample thickness was in the order of 1. So the scattering contribution from the unreacted GC core material could be subtracted linearly from the scattering curve, as exercised in the experimental part of this thesis.

Figure 6.66 displays the corrected data scattering curves of SIGRADUR®K GC with 60 microns nominal thickness, activated for various times at 450°C. An unactivated sample was measured as a reference and is displayed as well. The data were also corrected for a Q^{-4}-decay, therefore the curves have a slope of -4 in the log-log plot.

The extrapolated scattering intensity at zero angle for the micropores increases by a factor of around 4.5 within 3 hours of activation (Table 6.17). As the scattering at zero angle is $\Delta n_f^2 N V^2$, it may be concluded that the pore volume increases as well.

Also a sample activated only one minute at 450°C was measured, and it was found that the overall SAXS intensity of this sample (56±7 a.u.) was somewhat higher than the SAXS intensity of the non-activated sample (72±5 a.u.).

To get impression of the influence of activation on the SAXS intensity, the intensity of the non-activated sample was subtracted from the intensity of the sample activated one minute, without any corrections for the film thickness. The intensity difference was 15±4 a.u. . The active film thickness and the burn-off are neglected in this estimation, because the thickness changes due to film growth and burn-off are small after one minute of activation.

One may conclude that the activated surface of GC after one minute of activation has a scattering power of around 20% to 25% of a non-activated GC sample, or, in other words, the scattering power of 12 to 15 microns non-activated GC. This

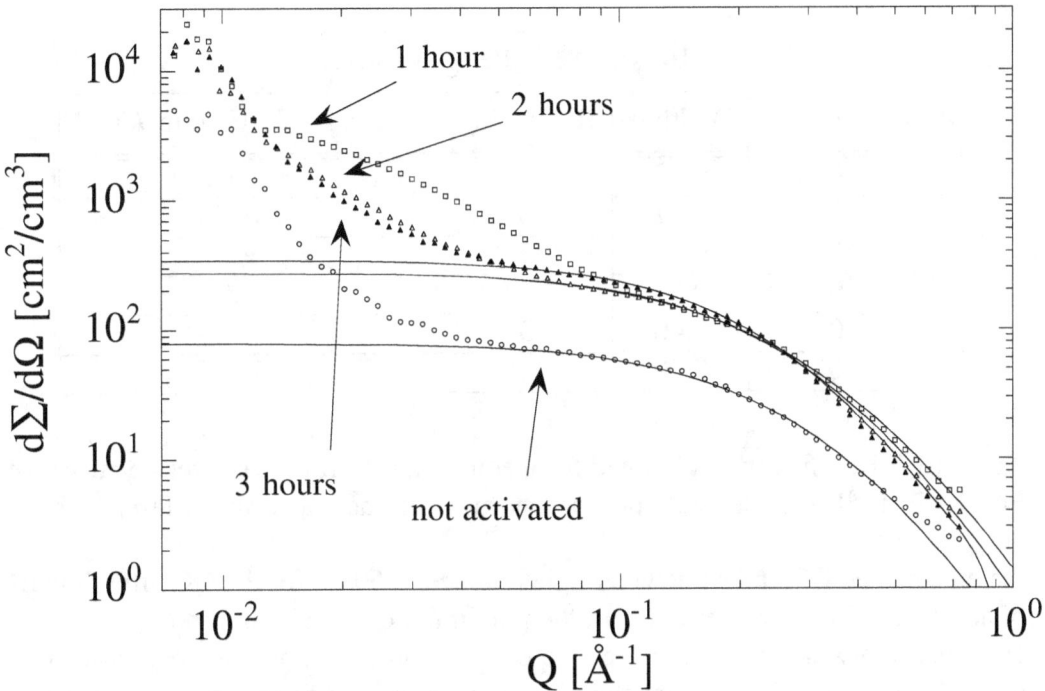

Figure 6.64: Log-log plot of SAXS curves for activated SIGRADUR®K with 60 microns thickness. Activation temperature was 450°C. The curve of an unactivated sample is displayed as a reference. The lines denote the fitted curve for the micropores. The curves of activated samples concern only the film properties, because the bulk contribution was subtracted.

6.6. SMALL ANGLE X-RAY SCATTERING ON GLASSY CARBON

result is particularly surprising because after one minute of activation the active film thickness should not increase 1 micron.

The scattering curves were fitted assuming logarithmically normal distributed spherelike pores, as described before in this thesis.
σ, R_0, and the scattering intensity at zero angle for the micropores were fitted. Results are displayed in Table 6.17. We obtain for the pore radius R_0 3.12 Å for

Micropore Fitting Results			
Activation time	Width σ	Radius R_0 [Å]	$\Delta n_f^2 \cdot N \cdot V^2$ [cm^2/cm^3]
0 min	0.45	3.12	3.62
30 min	0.37	3.81	10.14
60 min	0.38	3.74	9.64
90 min	0.38	3.97	9.66
120 min	0.40	3.91	11.74
180 min	0.40	4.50	16.05

Table 6.17: Best fit values obtained from equation 5.16 for the scattering curves of SIGRADUR®K samples with 60 micron thickness, activated for different times.

the unactivated GC and somewhat higher values (4.50 Å) for the GC after 3 hours of activation. These values concern the maximum of the distribution.
However, the width σ of the pore size distribution does not change significantly during activation, which means that all micropores are affected to the same extent when the film grows and insofar the pore growth is a homogeneous process.
Using the fitted values for R_0 and σ, it is possible to calculate and plot the pore size distribution. The distribution functions $P(R)$ for the pore radii in the active film and bulk material of three different samples are plotted in Fig. 6.67.
The pore radius R_0 enlarges upon activation from 3.12 Å (not activated) to 4.50 Å (3 hours activated), while the width σ remains constant around a value of 0.4 [20].
The scattering curves of samples activated for 30 and 60 minutes could not be fitted with sufficient accuracy. This was because the micropore part of the scattering curves of the samples activated for 30 minutes and 60 minutes is overlapped from additional structures, which occur during oxidation (see log-log plot in Figure 6.66).

---
[20]These data were confirmed by Dr. H.-G. Haubold, IFF, Forschungszentrum Jülich GmbH, who analyzed the scattering curve of one non-activated and one fully activated GC sample with the JUSIFA SAXS evaluation software.

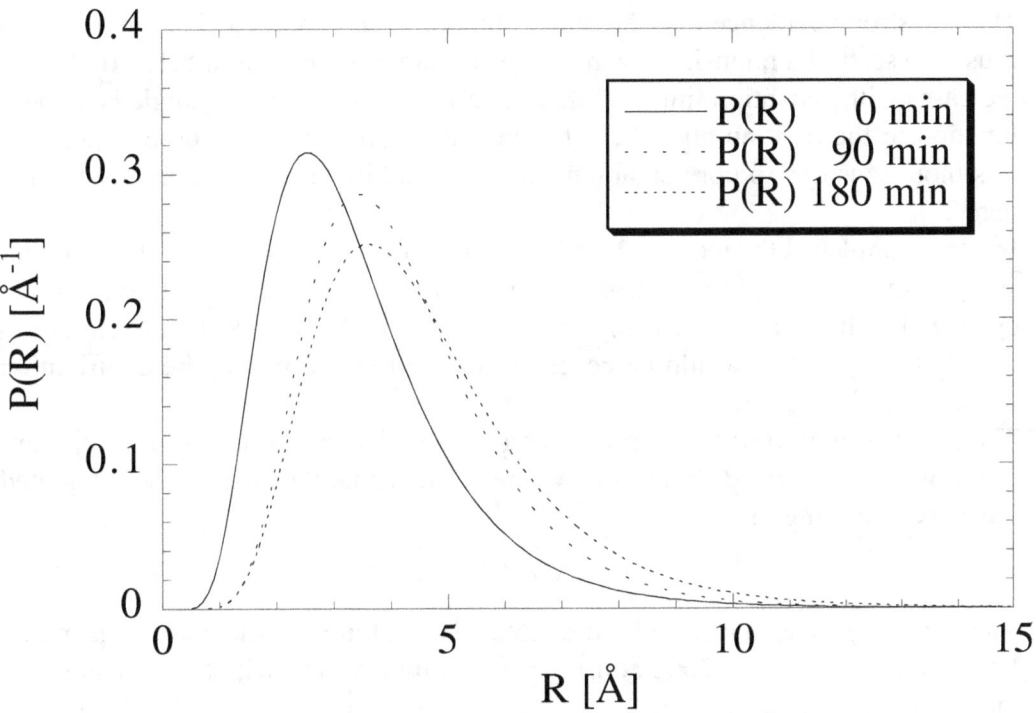

Figure 6.65: Pore size distribution for non-activated and for activated GC. Samples were activated at 450°C. The maximum of the distribution shifts towards somewhat larger radii during activation.

6.6. SMALL ANGLE X-RAY SCATTERING ON GLASSY CARBON

Throughout the Q-range a remarkable increase of the scattered intensity is found during the first 60 minutes of activation (Figure 6.66). The increase is particularly pronounced in the Q-range from 10^{-2} to 10^{-1} Å^{-1} and is proportional to $Q^{-1.5}$. The shape of the objects which account for this slope must be quite prolate (Q^{-1} corresponds to cylindrical pores).

The additional structures arise very probably from the evolution of the pore network. For further evaluation of these structures, the fitted curves for the micropore contribution have to be subtracted from the data curve. However, this was beyond the scope of this work.

There is strong evidence that the pore volume increases with activation time, because the scattered intensity at zero angle (Figure 6.66, Equation 5.17, Table 6.6) increases with activation time. As the scattering intensity at zero angle is a linear function of the pore number N, but a quadratic function of the pore volume V, it is more sensitive to pore enlargement than to an increase of the number of the pores.

For the samples activated for 2 and 3 hours, the region of the scattered intensity proportional to $Q^{-1.5}$ is shifted towards smaller Q-values, which reflects that cylinder-like pores are increasing during activation. For a more detailed interpretation of this result it would be necessary to subtract the micropore contribution of the scattering curve.

The porosity p or void fraction v of non activated samples and of the film contribution from activated samples ($p = 100\,v$) was calculated from the integrated intensity according to

$$v = N \cdot V / V_{sample}. \tag{6.66}$$

The porosity p is displayed in Figure 6.68. The evolution of the porosity p arising from pores smaller than 200 Å during activation measured with SAXS can be divided into three regions:

(i) During the first 30 minutes of activation, the porosity increases drastically from 7% to 28%.

Samples activated for short times ($t \leq 10$ minutes) do not have a film with a significant thickness, and data obtained from these samples mainly reflect properties of the interface between bulk and film, which is the chemical reaction frontier and which has a pronounced porosity.

(ii) Upon further activation the film becomes thicker, while the thickness of the interface remains constant and the scattered intensity is governed more and more by the active film.

Therefore, between 30 and 90 minutes of activation, p decreases linearly to around 21% and, then (iii), remains nearly constant upon further activation.

While the porosity of unactivated GC measured with SAXS is around 7%, the porosity of the same sample determined by XRD is around 23%. The difference

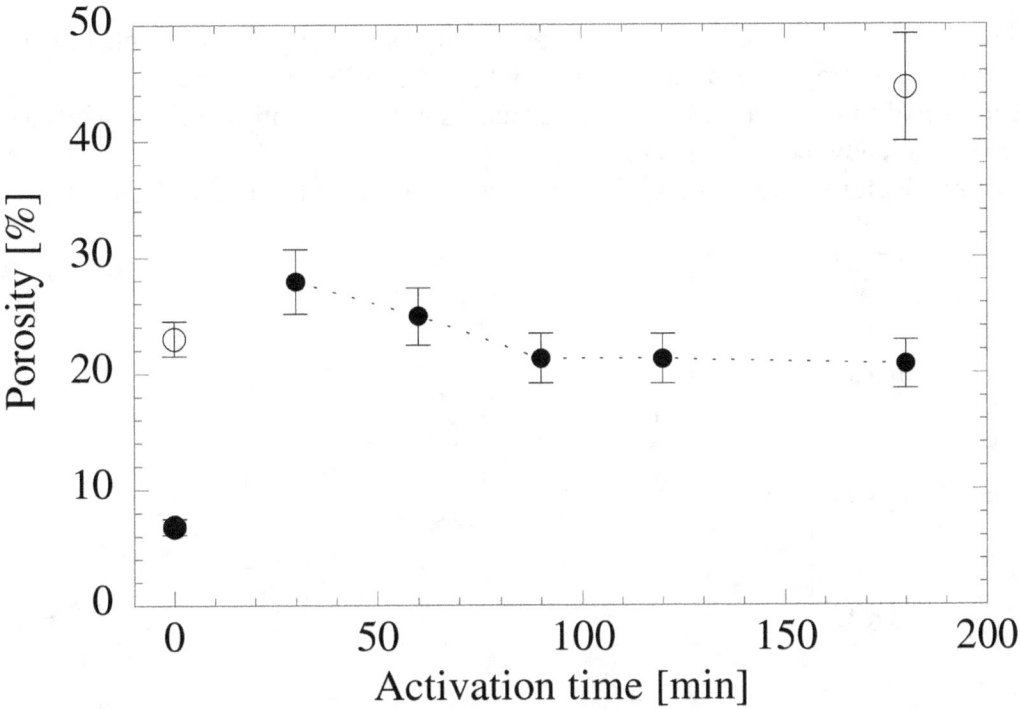

Figure 6.66: Porosity of SIGRADUR®K with 60 microns thickness vs. activation time. Open symbols concern data as found by XRD (X-ray density ρ_x). Filled symbols concern data found by the integrated SAXS intensity. While the XRD porosity of the non activated sample is around 23%, the SAXS porosity of the same sample is only around 7%. During 30 minutes of activation the SAXS porosity increases to around 30%, however.

of the porosity values from XRD and from SAXS could arise from the fact that XRD determines the X-ray density from the position of the (002)-peak and the apparent density of GC only. Amorphous carbon with slightly different density will not alter the evaluated porosity very much.

On the other hand, the porosity as obtained from SAXS does not include pores larger than 200 Å. Therefore the porosity obtained from XRD is close to the total porosity of the whole sample.

The Porod plots showed a nonzero slope for large scattering vectors, probably due to the fractal properties of the GC material [25, 75, 126]. Therefore the scattering curves had to be corrected by a background subtraction until the slope vanished [67], as already mentioned before.

The results for the volumetric surface areas are displayed in Figure 6.69. The in-

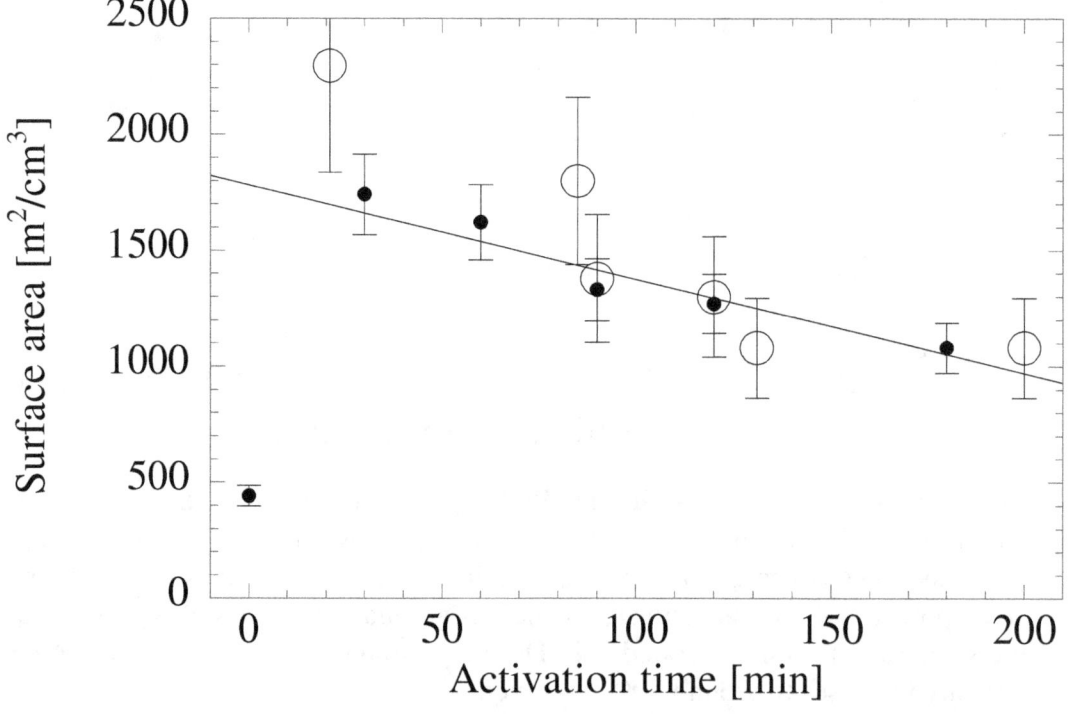

Figure 6.67: Evolution of the internal surface area during activation. The closed symbols concern data found by SAXS (Porod plot). The open symbols concern data found with nitrogen gas adsorption (BET). The straight line is obtained by the linear least square fit of SAXS data ($t¿0$) and BET data ($t¿90$), but should serve more as a guide to the eye.

ternal surface area of unactivated GC (SAXS) is only around 450 m^{2}cm^{-3}. The surface area of the activated GC samples obtained by SAXS decreases with activation time. The sample activated for 30 minutes has a surface area of nearly 1750 m^{2}cm^{-3}, whereas the surface area of the sample activated for 3 hours is around 1050 m^{2}cm^{-3}.

Obviously the internal surface area of the GC increases drastically during the initial stage of activation by a factor of around 4. It is pointed out that also the scattering intensity at zero angle for the micropores increases by nearly the same factor during initial stage of activation.

The results of the BET measurements on the activated GC samples show the same trend of decreasing internal surface area upon activation and are in good agreement with the SAXS results.

The value of around 1750 m^{2}cm^{-3} for a volumetric surface area appears to be rather high. A linear fit of selected measured values (SAXS values of activated samples and BET values lower than 1800 m^{2}cm^{-3}) yields a rate of decrease for the surface area of around - 3.8 m^{2}cm^{-3}min^{-1} and an intercept of around 1780 m^{2}cm^{-3}.

This intercept can be interpreted as the internal surface area of activated GC, extrapolated to 0 minutes of activation, or the internal surface area of the interface between bulk and film. The main reason for this large value could be a very strong scattering power of the material [21].

Probably this interface should be better regarded as having a high microscopic roughness than having a large internal surface area.

The interface between bulk and film is present throughout the whole activation time, but not taken into account by the background subtraction (Experimental, Chapter 4.4). Therefore its contribution to the internal surface area accounts most for short activation times. The porosity contribution of pores smaller than 200 Å remains constant between 90 and 180 minutes activation, while the volumetric surface area decreases linearly.

From the pore volume V, as obtained from the integrated intensity Q_o, and the volumetric surface area A as obtained from the Porod-plot, a mean value $\overline{\langle R \rangle}$ for a sphere pore radius and its corresponding area $\overline{\langle A \rangle}$ and volume $\overline{\langle V \rangle}$ can be estimated as follows:

$$\overline{\langle R \rangle} = 3\frac{V}{A}, \quad \overline{\langle A \rangle} = 4\pi\overline{\langle R \rangle}^2, \quad \overline{\langle V \rangle} = 4/3\pi\overline{\langle R \rangle}^3. \tag{6.67}$$

The values found are displayed in Table 6.18. These values do not only concern the micropores, but all pores measured with SAXS. They can be regarded as a mean upper limit for the pore radii. For comparison, the mean micropore radii

---

[21]Note that the sample activated only one minute has a surface with a scattering power of around 12 to 15 microns thick non-activated GC.

6.6. SMALL ANGLE X-RAY SCATTERING ON GLASSY CARBON

Act. time	⟨R⟩ [Å]	⟨A⟩ [Å^2]	⟨V⟩ [Å^3]	$\overline{\langle R \rangle}$ [Å]	$\overline{\langle A \rangle}$ [Å^2]	$\overline{\langle V \rangle}$ [Å^3]
0 min	3.45	183	315	4.6	266	408
30 min	4.07	239	426	4.8	290	463
60 min	4.02	235	421	4.7	276	435
90 min	4.23	264	511	4.9	295	493
120 min	4.26	263	497	5.0	319	524
180 min	4.62	315	673	6.0	452	905

Table 6.18: Comparison of mean values for micropore radius, surface area and micropore volume, as obtained from micropore fitting (⟨R⟩, ⟨A⟩, ⟨V⟩) and Porod-plot plus integrated intensity ($\overline{\langle R \rangle}$, $\overline{\langle A \rangle}$, $\overline{\langle V \rangle}$), for different activation times.

and their mean internal surface areas are displayed in Table 6.18 as well.

For a specific activation time, the pore radii $\overline{\langle R \rangle}$ are larger than the pore radii as obtained from the micropore fitting, because somewhat larger pores which were not included from the fitting routine contribute to the internal surface area and pore volume.

This is also valid for the pore area $\overline{\langle A \rangle}$ and the pore volume $\overline{\langle V \rangle}$ of the single pore. There is no doubt that the pores which account for the very high volumetric surface area of the oxidized GC are very small, because their mean value $\overline{\langle R \rangle}$ is 6 Å for the pore radius.

The effect of pore enlargement upon activation is in line with the *falling cards model* [30] and with the results found in XRD. The pores of GC are regarded as being polyhedra bordered or enclosed by stacks of graphene sheets [25].

The activation process preferably burns off interlinking and less ordered carbon atoms between neighboring stacks and graphene sheets. The graphene sheets may then undergo some kind of rearrangement by coalizing with one of the nearest stacks. They no longer act as a separation and leave behind larger pores.

In this simple model, the pore volume of the porous body would remain constant if burn-off of material would be neglected, but the number of pores and the internal surface area of the porous body would decrease.

This behavior is in fact observed between 90 and 180 minutes of activation (Figures 6.68 and 6.69).

Additionally, in the falling cards model a pore growth seems more probable than creation of additional pores, because pore growth is related to coalescence of pores, thus the pore number decreases.

Some more information on the structure of the GC can be inferred by analyzing the scattering curves. The asymptotic behaviour of the scattering curves for large

Q values was studied, and it was found that the slope of the scattering curves does not equal -4, as would be expected for smooth and non-oriented voids [68, 69, 67]. The non activated GC showed a slope of 2.4, while the fully activated GC showed a slope of 3.0 (Table 6.19). In the case of surface fractals, the number D_s - 6 equals the slope of decay for large Q [127]. D_s is the surface fractal dimension.
For n = 2.4 a surface fractal does not exist. Therefore 2.4 must be interpreted as the volume fractal dimension: D=2.4.
For n = 3 there is a surface fractal, because D_s - 6 = 3 = n, and 3 is the limiting case for a surface fractal which extends throughout the whole particle volume. Therefore D_s = 3 must be interpreted as the surface fractal dimension of the voids and pores in the GC. Fractal structures often evolve during stochastic growth processes [127, 128]. Especially char and charcoals have a fractal dimension D greater than 2 [129, 130, 77].
As the fractal dimensions of the investigated GC are between 2.4 and 3.0, the system under investigation represents a *volume fractal*. As the film material after 3 hours of activation has a dimensionality of 3.0, which is also the euclidean dimensionality, the pores have to be regarded as compact homogeneous objects with a smooth surface [66].
The activated GC represents a non-trivial limit because the dimension of the volume fractal and the dimension of the surface fractal are 3. This corresponds to the case of a porous solid which is sufficiently compacted that both the mass and the internal surface become uniformly space filling [77].
Such solid should yield a scaling law of $\sim Q^{-3}$ as predicted by the theory of Wong and Bray [131, 132].
Therefore the internal surface and the mass are extended through the whole particle volume [77, 66].
It is pointed out that this is the unique and indispensable feature of the activated glassy carbon electrode for electrochemical double layer capacitors !
As was already described in Chapter 3, Section 1, the *Euler-Poincaré-Characteristic* ϵ of the pore space is decreasing during activation, because pores are getting interconnected. The SAXS results qualitatively and directly proof that activated GC consists of an extensive network of carbon particles and an extensive network of pores.
Table 6.19 lists the exponents of asymptotic decay for non-activated and variously activated SIGRADUR®K samples with 60 micron thickness. The activation temperature was 450°C. To overcome any doubts or uncertainty in the procedure of bulk contribution subtraction, also raw data curves were investigated. The exponent was determined for the overall sample (bulk+film) and also for the film only to justify whether the bulk subtraction alters n systematically.
The exponent of decay changes from 2.40±0.02 (non-activated GC) to 3.00±0.06 (3 hours activated film) during activation.

6.6. SMALL ANGLE X-RAY SCATTERING ON GLASSY CARBON

Activation time [min]	n with bulk-subtraction	n no bulk-subtraction
0	2.38	2.42
30	2.24	2.41
60	2.31	2.53
90	2.53	2.78
120	2.64	2.84
180	2.94	3.06

Table 6.19: Exponent n of decay for scattering curves of 60 μ SIGRADUR®K as a function of activation time. With proceeding activation the decay becomes more pronounced. No Porod Q^{-4}-behaviour is observed.

The exponent of the raw data is always slightly smaller than the exponent of corrected curves, but the values experience the same trend, i.e. the exponent increases with increasing activation time, apart from short activation times, where a minimum is observed.
For better view, data points are plotted in Figure 6.70.
Despite a small shift on abscissa the values for n show the same trend: for short activation, n decreases; for longer activation times n increases again, with the result that a minimum for n is obtained around 30 minutes of activation.
Also in Figure 6.70, the X-ray density of the same samples is displayed (filled circles, on left axis). The data points for ρ_x exhibit exactly the same trend with a minimum around 30 minutes of activation. The fact that the X-ray density has a minimum around 30 minutes means that either (i) the distance between ordered graphene sheets has a maximum or (ii) the graphene sheets are maximally disordered. Anyway, in this case the graphene sheets are most separated and can be regarded more likely as sole sheets than closed stacks. However, planar structures such as sheets exhibit an exponent of decay of n=2. In so far the stage of short activation must be regarded as a transition state.
Although the raw data were corrected (subtraction of a constant background to obtain the Porod behaviour), the Porod constant did not change. The GC samples therefore may be regarded as a class of *robust volume fractals* [133].

6.6.3 Consideration of Plausibility

Recall the pore diameter of GC which was determined to be around 11 Å (14.8 Å) for SIGRADUR®K (SIGRADUR®G) by the Guinier plot.
One may ask whether such values are physically possible at all. It follows an estimation whether the data found for GC are reasonable, based on simple geo-

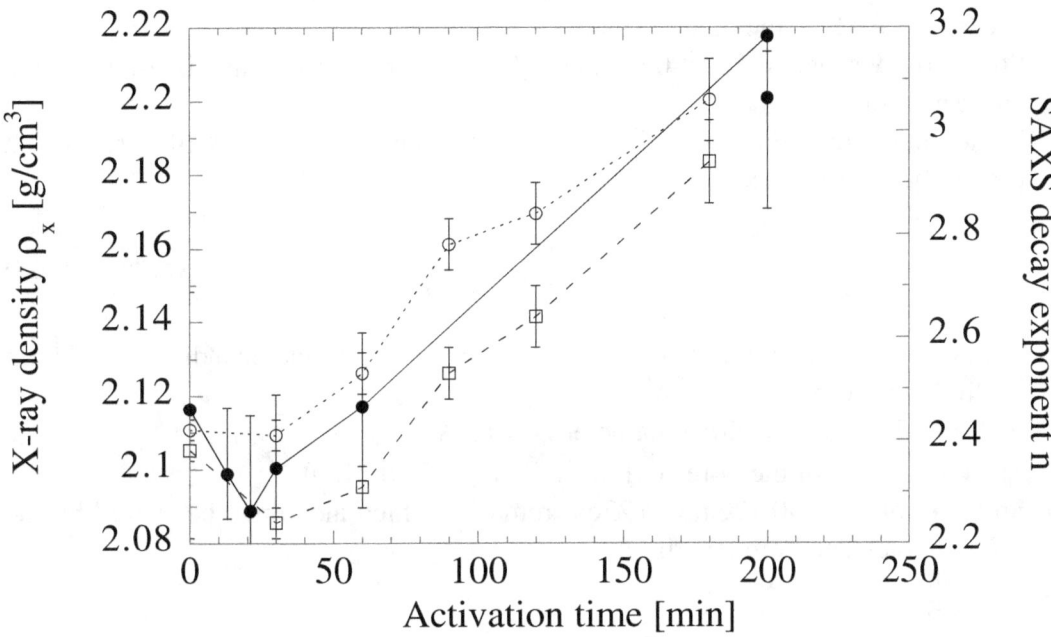

Figure 6.68: Comparison of the scattering contrast ρ_x (●) and exponent n of decay for the 60 micron thick SIGRADUR®K samples for various activation times. Open symbols denote data points for n. Open squares show n of scattering curves with no bulk subtraction. Data points after bulk subtraction represent solely film properties and are denoted with open circles. The activation temperature was 450°C.

6.6. SMALL ANGLE X-RAY SCATTERING ON GLASSY CARBON

metrical considerations.

Consider GC (ρ_{GC}=1.53 g/cm^3) being made from graphite ($\rho_{Graphite}$=2.26 g/cm^3) and from pores. Then, in a volume of 1 cm^3 GC, the graphite fills a space with 0.68 cm^3. Therefore 0.32 cm^3 of the volume is empty space.

For clarity, refer to the schematic in Figure 6.71. Although an oversimplification of the true situation in GC, the schematic may serve well to roughly estimate the pore size and internal surface area.

1.53 g of graphite fill a space of $\frac{1.53}{2.26}$ cm$^3 \approx 0.68 cm^3$. This is also valid for SIGRADUR®K with a mass density of 1.53.

So an empty space of 1cm^3-0.68 cm^3 = 0.32 cm^3 remains, which is used up by a number N of voids.

We assume a pore diameter of 11 Å and calculate the number N of pores which use up the empty space:

$$V = N \times \frac{4}{3}\pi R^3 \implies N = \frac{3 \cdot 0.32 cm^3}{4 \cdot \pi \cdot (5.5 \cdot 10^{-8} cm)^3} = 4.59 \cdot 10^{20} \text{voids}. \quad (6.68)$$

The N voids are distributed throughout a space of 1 cm^3, and on a distance of 1 cm we find therefore

$\sqrt[3]{N} = 7.7 \cdot 10^6$ voids with a diameter D = 11 Å.

The total length of the pores in line is $\sqrt[3]{N} \cdot D = 0.77 cm$.

So a gap of 1cm - 0.77cm = 0.23cm remains, which also must be shared by the 7.7·10^6 voids per 1 cm length:

$\frac{0.23 cm}{7.7 \cdot 10^6} \approx 3 \text{Å}.$

In one cubic centimeter SIGRADUR®K glassy carbon therefore are approximately 4.6·10^{20} voids with 11 Å diameter and 3 Å nearest neighbour distance.

The overall internal surface area of 4.6·10^{20} voids with 11 Å diameter yields 1750 m^2.

If different dense packed arrangements are considered (simple cubic, body centered cubic or face centered cubic), the corresponding void distances are 2 Å, 3.2 Å and 3.6 Å. In this model it was assumed that all voids have the same diameter. When different void diameter are taken into account, in conjunction with the theory that voids are built up by graphene sheets, even single graphene sheets, then the result that the pores have diameters of only 10 Å is reasonable.

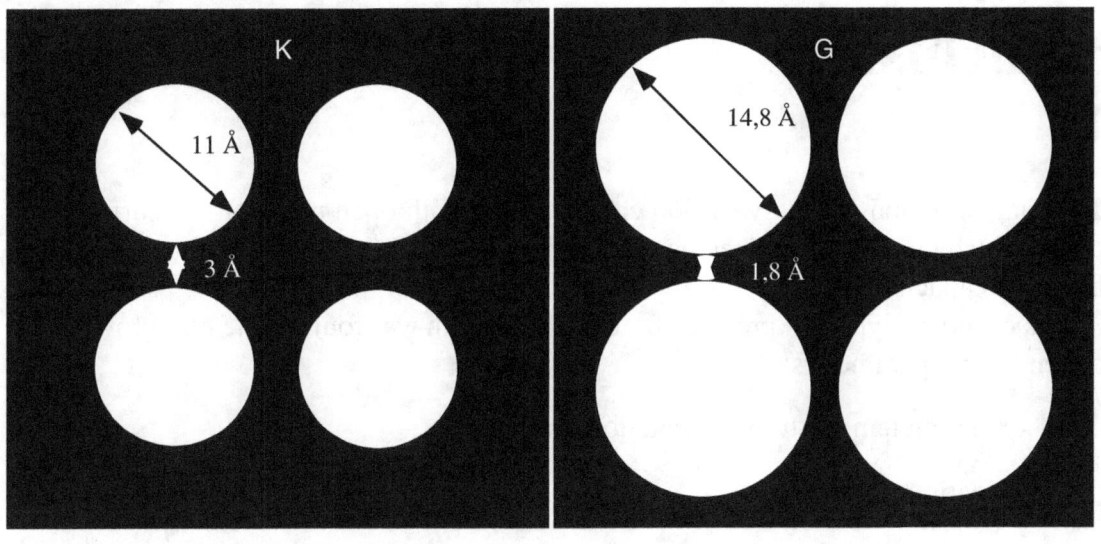

Figure 6.69: Schematic representation of voids in a GC matrix. The left part represents the SIGRADUR®K with 11 Å void diameter, the right part represents SIGRADUR®G with 14.8 Å void diameter.

Chapter 7

Conclusions

Thermal oxidation of glassy carbon creates a film with open pores on the surface of GC. Such GC can be utilized as an electrode material for electrochemical double layer capacitors.
An exact and analytical expression for the film growth was found. The film growth depends on 4 parameters:

- oxidation temperature ($\leftarrow$ reaction rate)
- oxidation time
- concentration of the oxidant
- structure of GC ($\rightarrow$ diffusion coefficient)

For large activation times, the film thickness yields a constant value (saturation film thickness), which is in principal the ratio of diffusion coefficient and reaction rate.
To achieve a thick saturation film thickness it is necessary to

1. oxidize at a low temperature (low reaction rate)
2. use a sort of GC with a large diffusion coefficient.

The saturation film thickness does not depend on the concentration of reactants. However, when the concentration of oxidant is increased, the saturation film thickness is reached in a shorter time.

While the influence of oxidation temperature and time and oxidant concentration is clear, the influence of the sort of GC activated is questionable.
The maximum achievable film thickness, which can be obtained by the proper choice of these 3 activation parameters, depends on whether the GC was pyrolyzed

at high or at low temperatures.

While the GC pyrolyzed at temperatures around 1000°C (SIGRADUR®K, including also K800) has the largest film thicknesses which could be measured (around 50 μm), the GC pyrolyzed at 2200°C (SIGRADUR®G) had nearly no active film at all (only around 1 μm) on its surface.

The active film thickness correlates both with the internal surface area and with the capacitance per apparent sample area.

These quantities correlate with the crystallographic structure of GC, as determined with XRD.

In particular, SIGRADUR®K has smaller crystallites and a higher defect density than SIGRADUR®G.

Therefore, the pyrolysis temperature of GC determines the microstructure of GC and consequently the film growth during thermal activation.

A thick film with a high energy density is obtained best with SIGRADUR®K, activated at not too high temperatures (400°C - 500°C). The energy density of the film material is 50 kJ/liter. The energy density of a 5 Volt capacitor stack with thermally activated GC electrodes was 0.6 kJ/liter.

Probably the microstructure is directly linked with the effective diffusion coefficient and in so far a key property for the activation of GC.

SIGRADUR®K therefore should have a better effective diffusion coefficient than SIGRADUR®G. The diffusion coefficient cannot be quantified easily. A high defect density could maintain that diffusion of reactant gases takes place so that an active film can grow, before it burns off.

Usually diffusion takes place in the vicinity of defects such as dislocations, grain boundaries or surfaces.

Direct proof that the pores in GC are opened after activation comes from adsorption measurements (BET gas adsorption, EIS and CV, SEM with $AgNO_3$ decoration).

As soon as the pores are opened and filled with electrolyte, an electrochemical double layer is built up so that it can be polarized for energy storage.

The internal surface area and the capacitance increase similar as the film thickness increases. The film obtained from K samples with 60 microns has open pores and a volumetric surface area between around 1800 m^2/cm^3 (30 minutes activation) and 1000 m^2/cm^3 (3 hours activation), as measured with SAXS and BET.

The values for the internal surface area of activated GC as measured with SAXS and BET agree fairly well with eachother. XRD and SAXS qualitatively reveal that the porosity of GC increases remarkably during thermal activation.

Quantitatively, however, the values for the porosity differ from eachother. At the initial stage of activation, only a part of the total surface area is wetted by the electrolyte, probably due to the fact that most pores are too small for the ions of the electrolyte, as can be seen from the micropore radii distribution.

However, the pores become wider during activation, as shown by SAXS, thus facilitating the penetration of electrolyte into the pores, and this might account for the increase of the volumetric capacitance upon activation as revealed by EIS.

A capacitance of around 120 F/cm^3 is achieved at potentials, where no chemical reactions occur. Theoretically, a capacitance of 400 F/cm^3 should be possible with an internal surface area of 2000 m^2/cm^3 and a double layer capacitance of 20 μF/cm^s. Considerably higher capacitances are physically not possible, because all pores which are large enough for the electrolyte to penetrate are filled with electrolyte.

The internal surface area can be increased, when GC with smaller pores is provided and the number of pores is increased. However, pores smaller than 7 to 9 Å do not contribute to the capacitance anymore. This lower limit for the pore size has to be obeyed.

SAXS measurements gave direct evidence that the pores grow during oxidation with the result that even a part of the smallest pores, which are originally to small for electrolyte ions to penetrate, become wide enough so that they can contribute to the capacitance. The overall growth and subsequent coalescence of pores causes a decreasing overall internal surface area.

During oxidation, in terms of the falling cards model, pores coalesce and grow, with the result that the porosity remains constant and the surface area decreases. However, as the pores are larger, the electrolyte may fill the pores and an electrochemical double layer can be formed [134, 135]. Thus the decreasing volumetric surface area and increasing volumetric capacitance can be explained.

It would be desirable to utilize glassy carbon with a somewhat narrower pore size distribution (sharp distribution around 10 Å diameter), such that as many pores as possible are just large enough for the electrolyte.

The capacitance depends also on the bias potential applied on the electrodes. Especially at 0.4 V there is additional capacitance arising from redox couples. The ratio of capacitance at 0.4 V and 0.9 V is around a factor of two. For a capacitor, these redox capacities are at 0.0 Volt, and one cannot benefit so much from the higher capacitance at the lower potential. A disadvantage of the use of aqueous electrolytes (here: sulfuric acid) is the limited potential window at which the capacitor can work.

For high power density applications it is mandatory to minimize all resistivities in the capacitor. The compact GC matrix allows for a fairly low materials resistance, and the sulfuric acid allows for the lowest possible electrolyte resistances at all. The diffusive resistance of the porous electrode was found to be critically high when GC was activated only short times. The minimum activation time for a low resistance was found to be around 1 hour.

Electrodes activated for times shorter than 40 minutes show a high specific diffusion resistance, maybe because only a small part of the porosity of the film with pores of sufficient size can be utilized by the electrolyte.

This result seems surprising, because the volumetric surface area (BET, SAXS) of the samples decreases at the same time. Obviously, the pores which are accessible to the electrolyte and account for the capacitance must be larger in size than most of the pores detected with gaseous adsorbents like N_2, which account for the high internal surface area.

There were hints that the large diffusive resistance could have its origin in a poor developed pore shape or pore opening.

In contrast to electrochemically activated GC, the thermally activated GC electrodes need no subsequent reduction after activation. This is a valuable feature of the thermal activation with respect to industrial manufacturing processes, because every additional processing step increases the production costs.

Appendix A

Activation of Glassy Carbon Powder

A disadvantage of the monolithic approach for the capacitor design is that the active film thickness is limited, either due to the competition between film growth and burn-off, or due to the peeling-off of cracked active film material from the GC sheets and disks.
The energy density of capacitors designed like the bipolar monolithic cell assembly is limited in this respect.
Activated glassy carbon in powder form permits that one can use any quantities of active material. Powder even can be brought into any shape, so that one is not restricted to rigid shapes such as sheets and disks.
Therefore the thermal activation of GC powder was studied also [136].

A.1 Experimental

Glassy Carbon powder was purchased from HTW Hochtemperatur-Werkstoffe GmbH, both types SIGRADUR®K and SIGRADUR®G.
The manufacturer provided us with a certification of the particle volume fraction X vs. particle radii (a numeric data table). The data were plotted and are displayed in Figure A.1. The particles were declared to have spherelike shape and diameter between 0.4 and 12 microns (G: 1 - 4 microns), which could be confirmed for the SIGRADUR®K in our laboratory with an optical microscope.
The SIGRADUR®G spheres are in majority smaller than the SIGRADUR®K spheres, because their maximum of the distribution is 4 μm for SIGRADUR®G and 12 μm for SIGRADUR®K.
For further specific evaluation of data it is necessary to know the volume fraction (Figure A.1) and also the distribution of the particle size, i.e. the particle radii fraction Y_i, as shown in Figure A.2 for SIGRADUR®K. The conversion from X

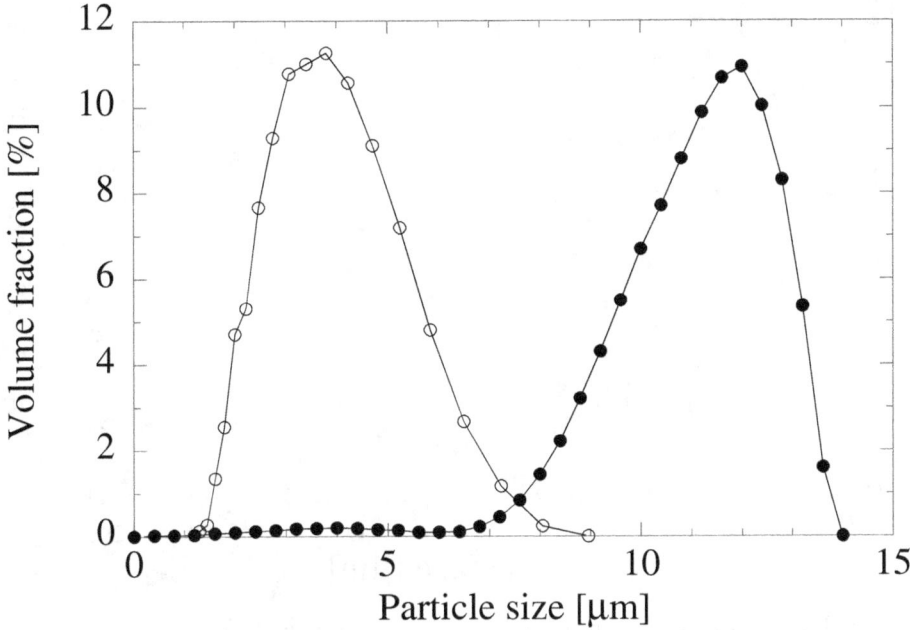

Figure A.1: Distribution of volume fraction X_i of GC powder SIGRADUR®K (•) and SIGRADUR®G (o).

to Y is shown below:

$$X_i = \frac{V_i}{V} = \frac{N_i \cdot \frac{4}{3}\pi R_i^3}{V}, \qquad (A.1)$$

V being the volume of all N particles in a certain amount of the powder, N_i of them having the radius R_i, and V_i therefore being the fraction of the volume arising from particles with radius R_i, thus $V = \sum_i V_i$ and

$$Y_i = \frac{N_i}{N} = \frac{V}{N}\frac{X_i}{\frac{4}{3}\pi R_i^3}. \qquad (A.2)$$

The results of the conversion are displayed in Figure A.2.

A.1.1 Oxidation of GC Powder

GC powder was thermally oxidized in the following manner: The non-activated GC powder was suspended in ethanol and then poured in a ceramic Petri dish of 12 cm diameter. The Petri dish was shaken by hand in order to cover the Petri dish walls with a thin film of the GC-ethanol suspension.
Within a few seconds, the ethanol evaporated and a thin film of GC powder was

A.2. EVOLUTION OF SURFACE AREA DURING ACTIVATION 219

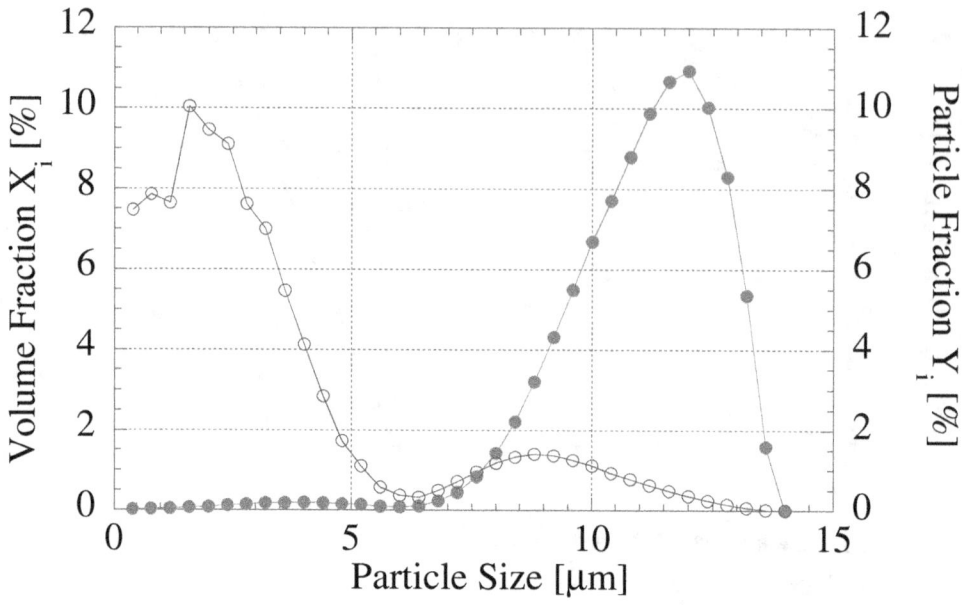

Figure A.2: Distribution Y_i of particle diameter fraction (left axis, open symbols) and distribution X_i of the volume fraction (right axis, closed symbols) of GC powder SIGRADUR®K.

left behind on the ceramic Petri dish. Care was taken to avoid agglomerates of GC in order to maintain that as many free particles as possible spread over the dish surface so that every powder particle had contact to air or oxygen, respectively.
The Petri dish then was brought into a preheated furnace, which had a temperature of 450°C. After a specified time, the Petri dish with the oxidized GC was removed from the furnace. Later, a second dish with non-activated GC powder was brought into the furnace to be activated for an other time, and so on. So a set of powder samples, all activated at 450°C could be provided. The relative weight loss $m(t)/m_0$ of the SIGRADUR®K powder was determined with a high precision laboratory micro balance. Gas adsorption measurements were carried out with N_2 on a Micromeritics ASAP 2000 test station. From the isothermal plots of V vs. p, the BET surface area of the powder in m^2/g was determined.

A.2 Evolution of Surface Area during Activation

A model for the evolution of the internal surface area of Glassy Carbon (GC) powder during thermochemical oxidation is established [112].
Results found by applying the model are compared with experimental results. The

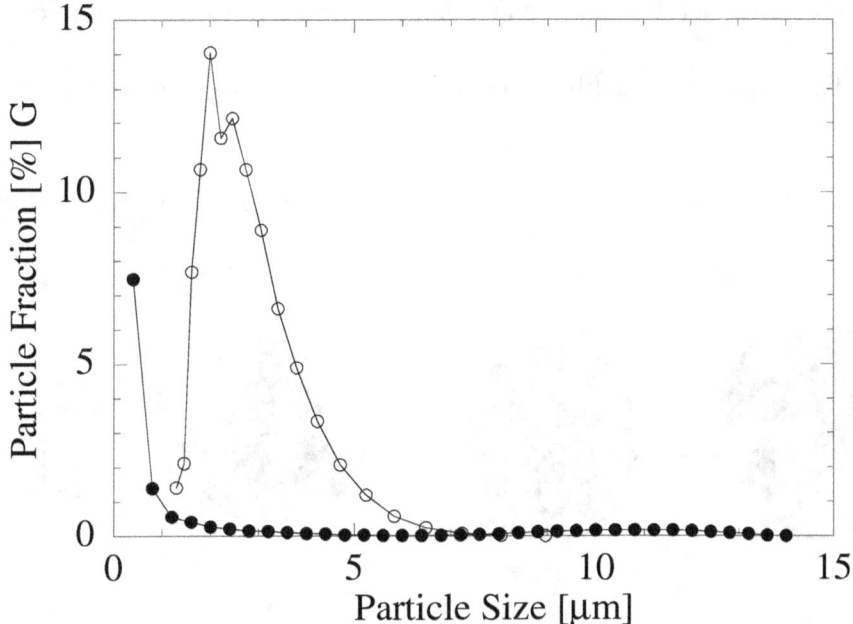

Figure A.3: Distribution of particle diameter Y_i of GC powder SIGRADUR®K (•) and SIGRADUR®G (◦).

internal surface area of thermochemically oxidized Glassy Carbon (GC) powder was measured and compared with results found by applying a model for the evolution of the internal surface of polydisperse GC powder upon thermal oxidation.

A.2.1 Introduction of the Model

From investigations on GC sheets and disks we know that a film with open porosity can be grown on the surface of GC.
The samples have a sandwichlike structure with an unactivated core with closed pores, which is enveloped by the film with open pores. We found that the thickness of the active film can be approximated by a square-root law. The actual thickness is described, however, by the $\mathcal{G}$-function.
The thickness of the whole sample decreases linearly.
The evolution of the internal surface area arising from open pores, which are created only by oxidation, can be monitored by gas adsorption measurements.
These results are taken into account for the model.
It is pointed out that the film growth in *spherical* particles is neither exactly described by the $\mathcal{G}$-function nor by a square-root function.
However, an exact solution for the film growth problem of the spheres is presented

A.2. EVOLUTION OF SURFACE AREA DURING ACTIVATION

in Appendix B to this thesis. Also an analytical solution for the film thickness is provided.
For simplicity, however, the film growth is approximated by a squareroot function here.

The oxidation of a spherical GC particle can be divided into several stages as illustrated in Figure A.4:

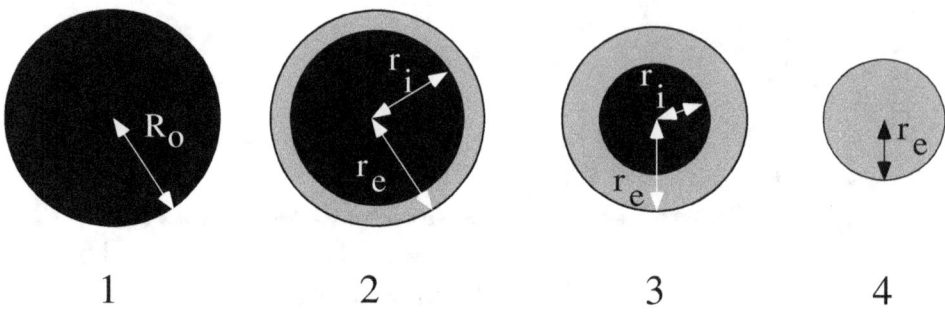

Figure A.4: Illustration of changes in particle radii during oxidation. The unreacted-core shrinking (decrease of black core) and the shrinking particle size (decrease of grey film) are important features of the model.

1. Initially there is a non-activated GC sphere with radius R_o and mass density ρ_i.

2. Then, upon oxidation, an outer shell with radius $r_e(t)$ of active film material with mass density ρ_e is created, and the active volume envelopes a non activated core with radius $r_i(t)$, which contributes to the weight of the powder, but not to the internal surface area. The thickness of the shell is $d_{film}(t) = r_e(t) - r_i(t)$, and $r_e(t) \approx R_o$.

3. Meanwhile a burn-off of the outer sphere occurs, and $r_e(t)$ decreases. Outside the sphere there is already a burn-off of active material, while the active film is growing deeper in the interior of the GC.

4. After a specific time there is no unactivated core GC anymore, but only active material.
   From now on only the fully activated GC sphere with radius $r_i(t)$ is burning off. The ratio
   $$\Omega = \frac{Internal\ Surface\ Area}{Mass}$$

remains constant, but the amount of material decreases [1].

The radii $r_e(t)$ and $r_i(t)$ change upon oxidation and are therefore time dependent. For a growth of active film, it is necessary that the rate of growing of the active film is larger than the rate of burning off the sphere.
The internal surface area vs. activation time as determined with BET can be calculated applying this simple model.

The mass of the partially activated sphere consists of the mass of the unreacted core with mass density ρ_i and radius $r_i(t)$

$$m_i(t) = \rho_i \frac{4}{3}\pi r_i^3(t) \tag{A.3}$$

and of the surrounding film with a mass density ρ_e and radius $r_e(t)$

$$m_e(t) = \rho_e \frac{4}{3}\pi (r_i^3(t) - r_e^3(t)). \tag{A.4}$$

The total surface area S of the particle, which can be covered by an adsorbent gas, consists of the geometrical surface area

$$S_{geom.}(t) = 4\pi r_e^2(t) \tag{A.5}$$

and the internal surface area of the active film volume

$$S_{int.}(t) = \Omega(r_e^3(t) - r_i^3(t)), \tag{A.6}$$

Ω being the internal surface area of the active volume in m^2/cm^3.
The internal surface area S related to the mass m equals:

$$S_{(t)}^* = \frac{S_{geom.} + S_{int.}}{m_i + m_e} = \frac{3r_e(t)^2 + \Omega(r_e(t)^3 - r_i(t)^3)}{r_i(t)^3(\rho_i - \rho_e) + r_e(t)^3 \cdot \rho_e}. \tag{A.7}$$

The function $S_{(t)}^*$ is piecewise defined because the radii vanish at some specific times $t_e = R_0/\beta$ and $t_i = R_0/(\beta + \gamma)$.
For an evaluation and comparison with experimental data, the distribution of the radii R_0, the mass densities and the time dependence of the radii and the maximum achievable BET surface area per gram must be known.

---

[1] It is an assumption that the volumetric internal surface area remains constant. This is not necessarily true, as was shown in subsection 6.6.2 and Figure 6.60. According to the falling cards model, the volumetric internal surface area may decrease as a result of structural changes upon oxidation. Structural changes occur in GC powder during oxidation [123].

A.2.2 Results and Discussion

As the distribution of the GC particles could not be measured in our laboratory, the manufacturers data were taken.

For the mass density of not-oxidized GC, values as reported from the manufacturer and as found in literature [16] were taken (SIGRADUR®K: ρ_i=1.53 g/cm^3; SIGRADUR®G: ρ_i=1.49 g/cm^3).

Mass densities for the active film of GC were not available with the exception of GC sheets SIGRADUR®K with 60 microns thickness, $\rho_i \approx$ 1.2 g/cm^3.

The internal surface area (BET) of the activated GC per active film volume was assumed to be 900 m^2/g for the SIGRADUR®K powder (G: 200 m^2/g), because the measured data tend to this value for long activation times (asymptotic behaviour). Additionally, from investigations on the 1 mm GC disks and 60 micron sheets we know that their active films have a BET surface area of around 900 m^2/g for large activation times.

X-ray Diffraction

Figure A.5 displays XRD diffractograms of the two kinds of GC powders studied (not activated).

The diffractograms reveal that structural differences in the SIGRADUR®G powder and SIGRADUR®K powder are similar to the structural differences in the monolithic plates.

The curve with the lower intensity is from SIGRADUR®K powder; the curve with the higher intensity is from SIGRADUR®G powder.

The large difference in intensity reveals that the SIGRADUR®G powder has a higher stage of crystallinity than the SIGRADUR®K powder, which was already found for the sheet and disk samples of SIGRADUR®K and SIGRADUR®G (Section 6.5.1., Figure 6.45.).

This result is in line with the findings for the burn-off, because the SIGRADUR®K plates burn-off in a shorter time (or at lower temperature) than the SIGRADUR®G plates.

The according diffractograms of 1 mm SIGRADUR®K and SIGRADUR®G samples are found in Figure A.6.

Also the SIGRADUR®K powder burns off faster (in shorter time at the same temperature) than the SIGRADUR®G powder.

Burn-off

From experiments on oxidized GC sheets of different types we know [103, 88, 137] that the active film thickness d_{film} follows approximately a squareroot-like

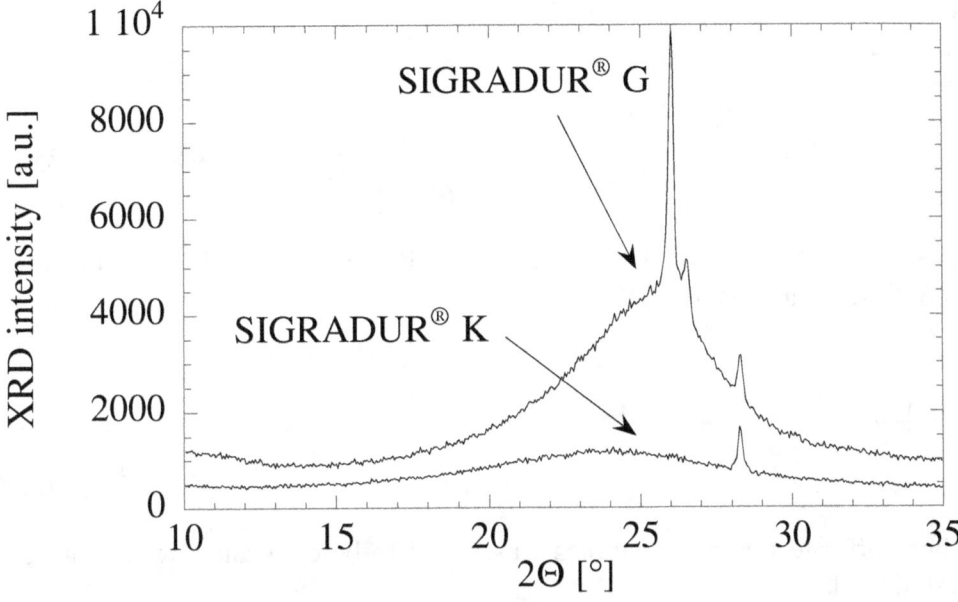

Figure A.5: X-ray diffractogramms of GC powder SIGRADUR®K (lower curve) and SIGRADUR®G (upper curve). Peaks around 28° are from corundum in the grain mill. The peak around 26° in the diffractogram of SIGRADUR®G could originate from highly crystalline graphite particles.

A.2. EVOLUTION OF SURFACE AREA DURING ACTIVATION

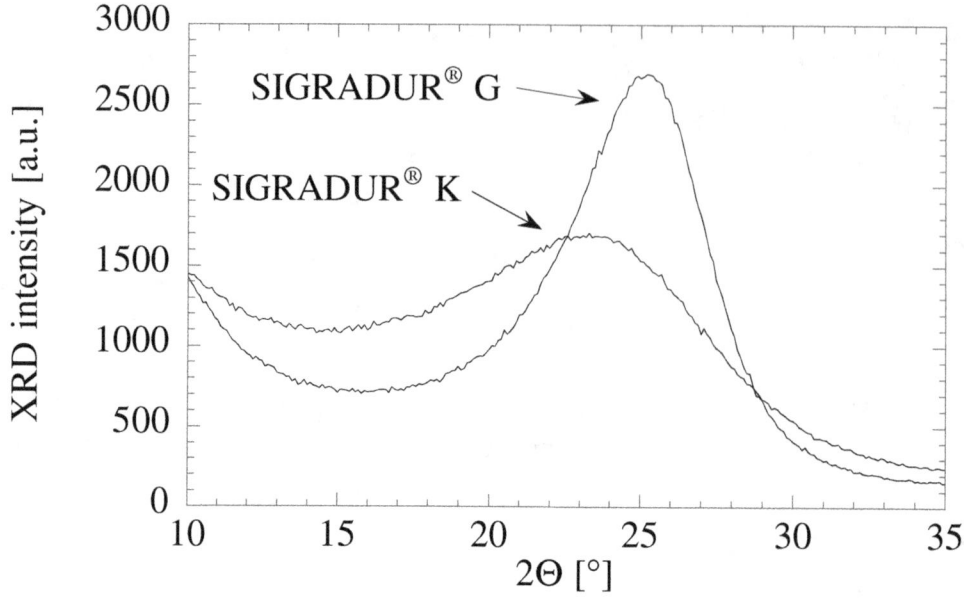

Figure A.6: X-ray diffractogramms of GC plates of 1 mm thickness SIGRADUR®K and SIGRADUR®G.

growth law:

$$d = \gamma \cdot \sqrt{t}. \qquad (A.8)$$

The burn-off rate for these samples was found to be constant. Results are displayed in Table A.1.

In our model we therefore assume also for the GC powder (i) a rate of the active film growth which is proportional to a squareroot like law of oxidation time [2] and (ii) a constant rate for the burn-off of GC material with

$$r_e(t) = R_0 - \beta t \geq 0, \quad r_i(t) = R_0 - \beta t - \gamma t^{1/2} \geq 0.$$

We point out that the growth rates and burn-off rates in Table A.1 deviate considerably from eachother, regarding different types of GC.

Unlike the weight loss due to burn-off of the large and flat sheet and disk GC samples, which was found to be linear, the weight loss of the spread out powder in the Petri dishes was found to decrease with the third power of sphere radius, which is in line with expectations from theory for spherical particles.

The relative weight loss of the SIGRADUR®K powder during oxidation is displayed in Figure A.7.

---

[2] This is actually not the right dependence, but a rough approximation.

APPENDIX A. ACTIVATION OF GLASSY CARBON POWDER

Burn-off rates and growth rates			
Temperature [°C]	Thickness [μm]	Growth γ [μm/$\sqrt{h}$]	Burn-off β [μm/h]
550	1000	-	197
500	1000	18.5	79
450	1000	38.7	22
450	100	11.8	-
450	60	11.0	3.7
450	K powder ($\approx 12~\mu$m)	-	2.2 ± 0.6
450	G powder ($\approx 5~\mu$m)	-	0.83 ± 0.3

Table A.1: Burn-off rates β and active film growth rates γ for various GC sheets (SIGRADUR®K) of different thickness and GC powder.

The experimental data could be fitted with an expression of the type

$$m(t) = m_0 \cdot \left(1 - \frac{\beta \cdot t}{R_0}\right)^3. \qquad (A.9)$$

For the radius of the SIGRADUR®K powder spheres, a value of $R_0 = 6.8~\mu$m, and for the burn-off rate a value of 1.6 μm/hour was obtained. The SIGRADUR®K powder particle radius is in good agreement with the particle size distribution as specified by the manufacturer (Figures A.1 - A.3).
During our measurements we also found that after around 150 minutes no SIGRADUR®K powder was visible in the furnace anymore, when it had been well spread over the Petri dish surface.
This means that the largest SIGRADUR®K powder particles, which have a diameter of around 14 microns, vanish after this time.
The burn-off rate for SIGRADUR®K powder obtained by this consideration was 7μ/150 minutes = 2.8 μ/hour.
Therefore the burn-off rate for SIGRADUR®K powder can be assumed to be between 1 and 3 μm/hour.
Similar considerations and observations for the SIGRADUR®G powder yield a burn-off rate of around 0.83 ± 0.3 microns/hour.
The burn-off rate of 60 micron GC sheets as mentioned before was found to be around 3.7 μm/hour.

BET Surface Area

BET measurements of the activated powder showed that the internal surface area per gram increases with activation time.

A.2. EVOLUTION OF SURFACE AREA DURING ACTIVATION

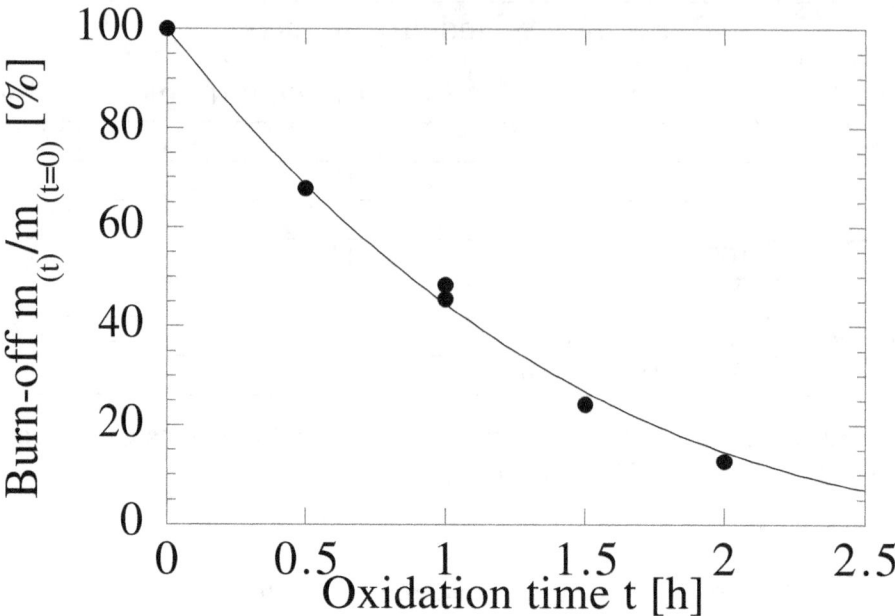

Figure A.7: Relative burn-off in [%] of SIGRADUR®K powder at 450°C. The drawn line represents the best fit according to equation A.9.

If the film of powder in the Petri dish before activation is too thick, the internal surface area of the oxidized powder is smaller than the saturation value, probably because GC in deeper regions near the dish walls is covered by surrounding GC and therefore protected against oxidation.

Results on the BET measurements are displayed in Figure A.8 and Figure A.10. For SIGRADUR®K an increase of the internal surface area from around 2 m^2/g at zero activation to around 900 m^2/g after two hours of activation is observed. A larger value for longer activation times is not expected, because the powder is consumed between 120 and 150 minutes of activation.
The value of 900 m^2/g is in line with results found on activated GC material, which was obtained after peel off from activated monolithic SIGRADUR®K.
After a specific time, the internal surface area remains at a constant value of saturation.
For SIGRADUR®G an increase of the internal surface area from around 2 m^2/g at zero activation to around 200 m^2/g after three hours of activation is observed.
The lower BET internal surface area for activated SIGRADUR®G powder is not surprising, because we know that the DLC of activated G-sheets and G-disks is also very low.

As no SAXS data of GC powder are available, also no direct information on the internal surface area of non-activated GC powder is available.

There should not be principal differences in the internal surface area of SIGRADUR®G and SIGRADUR®K powder, because their microstructure differs not much from the SIGRADUR®K and SIGRADUR®G sheet and disk samples.

One should therefore assume that micropore sizes and internal surface areas are similar between monolithic samples and powders.

Possibly the SIGRADUR®G powder has larger and less pores than SIGRADUR®K powder, as SIGRADUR®G sheets have larger pores than SIGRADUR®G sheets.

Fitting

Figure A.8 displays measured BET data and theoretical values concerning the SIGRADUR®K powder, as computed according to equation A.7 for three different sphere radii R_0. Note that equation A.7 only respects a monodisperse distri-

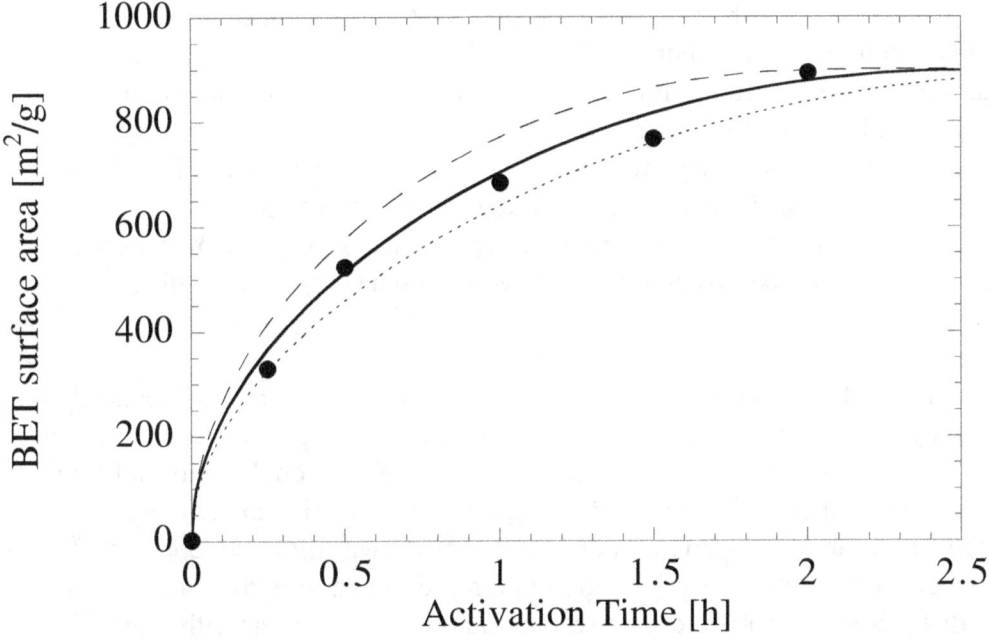

Figure A.8: Measured and calculated BET surface area for SIGRADUR®K powder. Three calculated curves are displayed with different assumed particle radii. Parameters: $\beta=0.87$; $\alpha=3.3$; $\omega=900$ for all curves. Upper dashed curve: $R_0=7$ μm; solid line: $R_0=8$ μm; lower dotted curve: $R_0=9$ μm.

bution of powder spheres.

A.2. EVOLUTION OF SURFACE AREA DURING ACTIVATION

Fitting parameters for SIGRADUR®G powder.				
Monodisperse		Bimodal Lognormal Distribution		
β	$9.16 \times 10E-5$	β	$0.15 \times 10E-4$	$0.15 \times 10E-4$
γ	$25.6 \times 10E-5$	γ	$0.15 \times 10E-4$	$0.15 \times 10E-4$
R_0	1.2764	R_0^a	$0.71 \times 10E-4$	R_0^b $1.26 \times 10E-4$
σ	$\rightarrow 0$	σ^a	0.115	σ^b 0.435

Table A.2: Fitting parameters for SIGRADUR®G powder. The left part denotes fitted values for the monodisperse distribution. The right part denotes fitted values for the bimodal lognormal distribution.

The best fit for monodisperse spheres is obtained with a radius of 8 μm and a burn-off rate of 0.87 and a film growth rate of 3.3.
The radius of 8 μm corresponds to 16 μm diameter, which matches fairly well with the data for the sphere radii, the largest diameter of which should be 12 microns (maximum of the distribution, Figure A.1).
The particle number distribution in Figure A.3 shows that a very large part of the particles is smaller than 4 microns.
Due to their early burn-off they are not present anymore after 1 hour of oxidation, so that they do not contribute anymore to the internal surface area.
From the technological point of view this means, that some part of material is wasted. This can only be avoided, when powder with a sharp particle distribution is utilized.

The BET data of the SIGRADUR®G powder were fitted with a monodisperse distribution (by applying equation (A.7) only) and also with 2 lognormal-distributions (bimodal) for the particle size (see Figure A.10) [3]. The bimodal lognormal fitted curve for the distribution is displayed in Figure A.9. The BET experimental data and fitted curves are displayed in Figure A.10. The drawn solid line represents the monodisperse distribution fit. The dotted line represents a fit with a bimodal lognormal distribution of sphere diameter. The burn-off and film growth rates were both 0.15×10^{-4} for β and γ. The other fit results are listed in Table A.2

---

[3] Similar routines as described in Chapter 5.2 and equations (5.15), (5.16), (5.18)-(5.20) were applied.

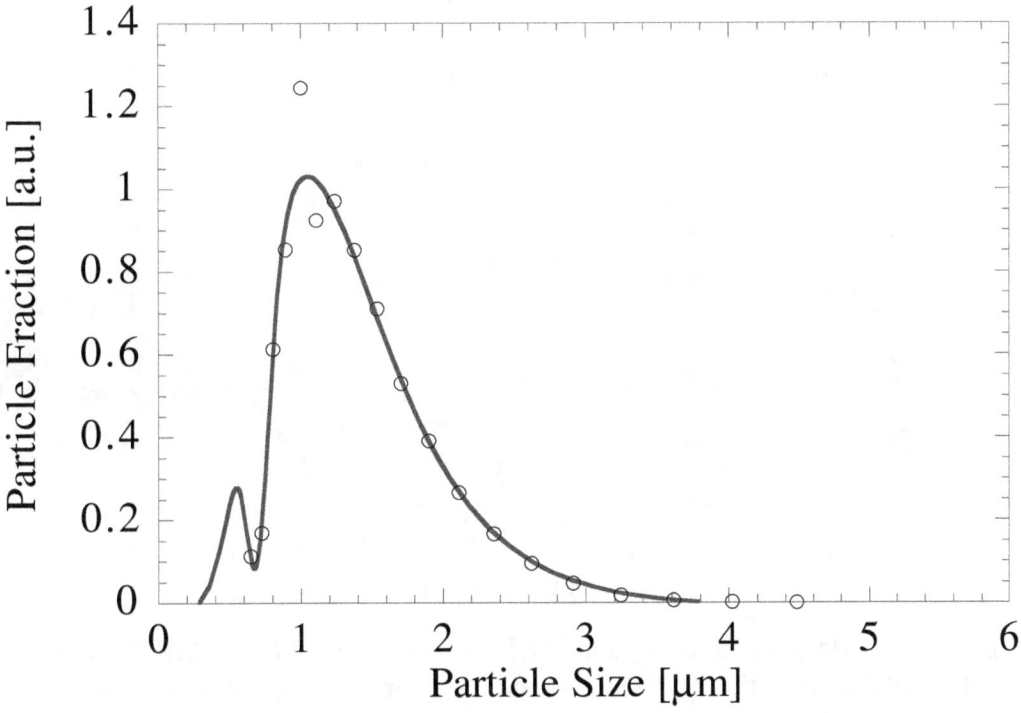

Figure A.9: Particle size distribution and fitted curve (bimodal lognormal distribution) for SIGRADUR®G powder. Particle diameter are written in μm.

A.2. EVOLUTION OF SURFACE AREA DURING ACTIVATION

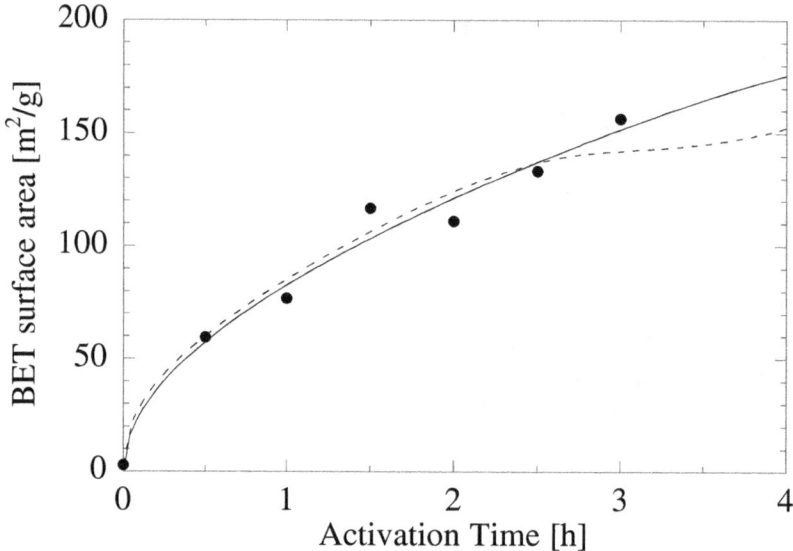

Figure A.10: Measured and calculated BET surface area of SIGRADUR®G powder. Dotted line: equation (A.7) and bimodal lognormal distribution applied. Solid line: equation (A.7) and monodisperse distribution applied.

Appendix B

Film Growth Model for Spherical Particles

When spherical particles are activated, the film growth model for the flat platelet samples can only be applied when the active film is very thin and when the sphere radius is very large.
In these cases, the solution for the plates is a good approximation.
When GC powder is oxidized, this approximation does not hold anymore because all particles become fully oxidized and no unreacted core remains finally, as illustrated in Figure B.1. The sphere geometry dominates the balancing of the reactions [138, 99] [1].

Therefore the model for the plates is extended to spheres. The solution of the differential equation is only presented as a sketch in this thesis. The complete solution requires a deeper algebraic treatment, which was beyond the scope of this thesis.
We start with the same considerations and equations as already used for the plate model in Chapter 6.2.
The only difference between the model for the flat samples and the model for the spheres is that in the former case the surface area S was constant, while for the latter case the surface area S is that of a sphere with variable radius R, which has to be taken into account accordingly for integration:

$$-\frac{dN_A}{dt} = 4\pi r^2 Q_A = 4\pi R^2 Q_{As} = 4\pi r_c^2 Q_{Ac} = constant \quad \text{(B.1)}$$

$$Q_A = D_e \frac{dC_A}{dr} \quad \text{(B.2)}$$

---

[1] W. Jander introduced 1927 a model for chemical reactions of a solid sphere with other reactants and neglected that the reaction surface of shrinking spheres is not constant, with the result that for long reaction times considerable deviations between theory and experiment occurred [99].

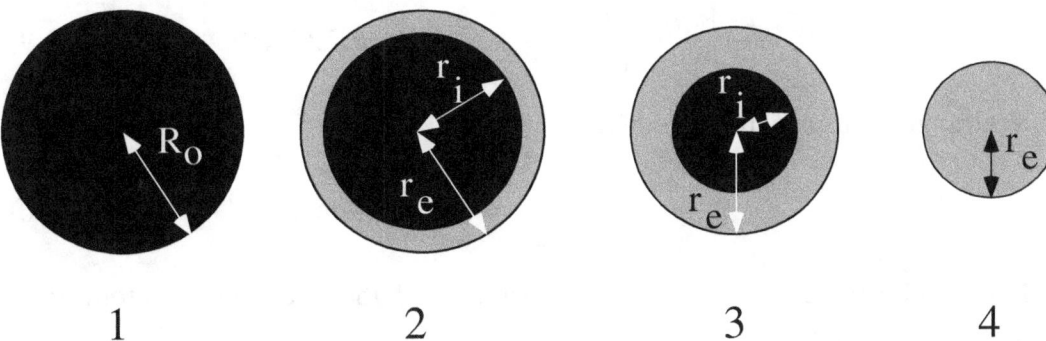

Figure B.1: Schematic representation of active film growth and of changes in pore radii during oxidation in GC spheres. The initial sphere radius before activation is R_0. The radius of the unreacted core at an arbitrary time is r_i, the radius of the overall sphere at the same time is r_e. After some specific time, no unreacted core is present anymore, and only active film material is obtained.

$$-\frac{dN_A}{dt} = 4\pi r^2 D_e \frac{dC_A}{dr} = constant \qquad (B.3)$$

$$-\frac{dN_a}{dt} \int_r^{r_c} \frac{dr}{r^2} = 4\pi D_e \int_{C_{Ag}=C_{As}}^{C_{Ac}=0} dC_A \qquad (B.4)$$

$$-\frac{dN_a}{dt}\left(\frac{1}{r_c} - \frac{1}{R}\right) = 4\pi D_e C_{Ag} \qquad (B.5)$$

This expression represents the conditions of a reacting particle at any time. From the diffusion through gas film we know that

$$-dN_B = -bdN_A = -\rho_B dV = -\rho_B d\left(\frac{4}{3}\pi r_c^3\right) = -4\pi \rho_B r_c^2 dr_c \qquad (B.6)$$

Therefore we find

$$-dN_A = \frac{-4\pi \rho_B}{b} r_i^2 dr_i \qquad (B.7)$$

For the radius of the unreacted core r_c and for the radius R of the whole sphere we write now

$$r_c = r_i, R = r_e = R_0 - \beta t \qquad (B.8)$$

$$-\frac{4\pi\rho_B}{b}\frac{r_i^2 dr_i}{dt}\left(\frac{1}{r_i}-\frac{1}{r_e}\right) = 4\pi\mathcal{D}_e C_{Ag} \tag{B.9}$$

Rearranging this equation, we find

$$r_i^2 dr_i\left(\frac{1}{r_i}-\frac{1}{r_e}\right) = -\frac{\mathcal{D}_e C_{Ag} b}{\rho_B}dt \tag{B.10}$$

Equation B.10 is parametrized with the parameter t, the reaction time. From equation B.8 we can write

$$dr_e = -\beta dt\,,\; dt = \frac{-dr_e}{\beta}\,. \tag{B.11}$$

In equation B.10 we can replace dt now and find the relationship between the radii r_i and r_e. So we can write a new ODE, having t not explicitly anymore as a parameter or variable, which describes the shrinking sphere and the film growth:

$$r_i^2 dr_i\left(\frac{1}{r_i}-\frac{1}{r_e}\right) = \frac{\mathcal{D}_e C_{Ag} b}{\rho_B \beta}dr_e \tag{B.12}$$

Several constants in equation B.12 can be replaced by one constant $\mathcal{C}$:

$$\frac{\mathcal{D}_e C_{Ag} b}{\rho_B \beta} = \mathcal{C} \tag{B.13}$$

So we can write a new equation, having t not explicitly anymore as a parameter or variable. Following equation is an ODE which describes the burn-off and film growth of the sphere particles:

$$\frac{dr_e}{dr_i} = \frac{1}{\mathcal{C}}r_i^2\left(\frac{1}{r_i}-\frac{1}{r_e}\right) \tag{B.14}$$

Equation B.14 will be expressed in a mathematically more general way: $x = r_i$, and $y = r_e$:

$$\frac{dy}{dx} = y' = \frac{1}{\mathcal{C}}x^2\left(\frac{1}{x}-\frac{1}{y}\right) \tag{B.15}$$

which is an ordinary differential equation (ODE) of the Abel type, 2nd type, class A.

No general procedure to solve this ODE is known. Also procedures as described in *Kamke* [139] could not solve this ODE exactly. However, within this thesis an exact solution of this Abelian ODE was found.

For the film thickness an analytical expression is derived. Figure B.2 displays the

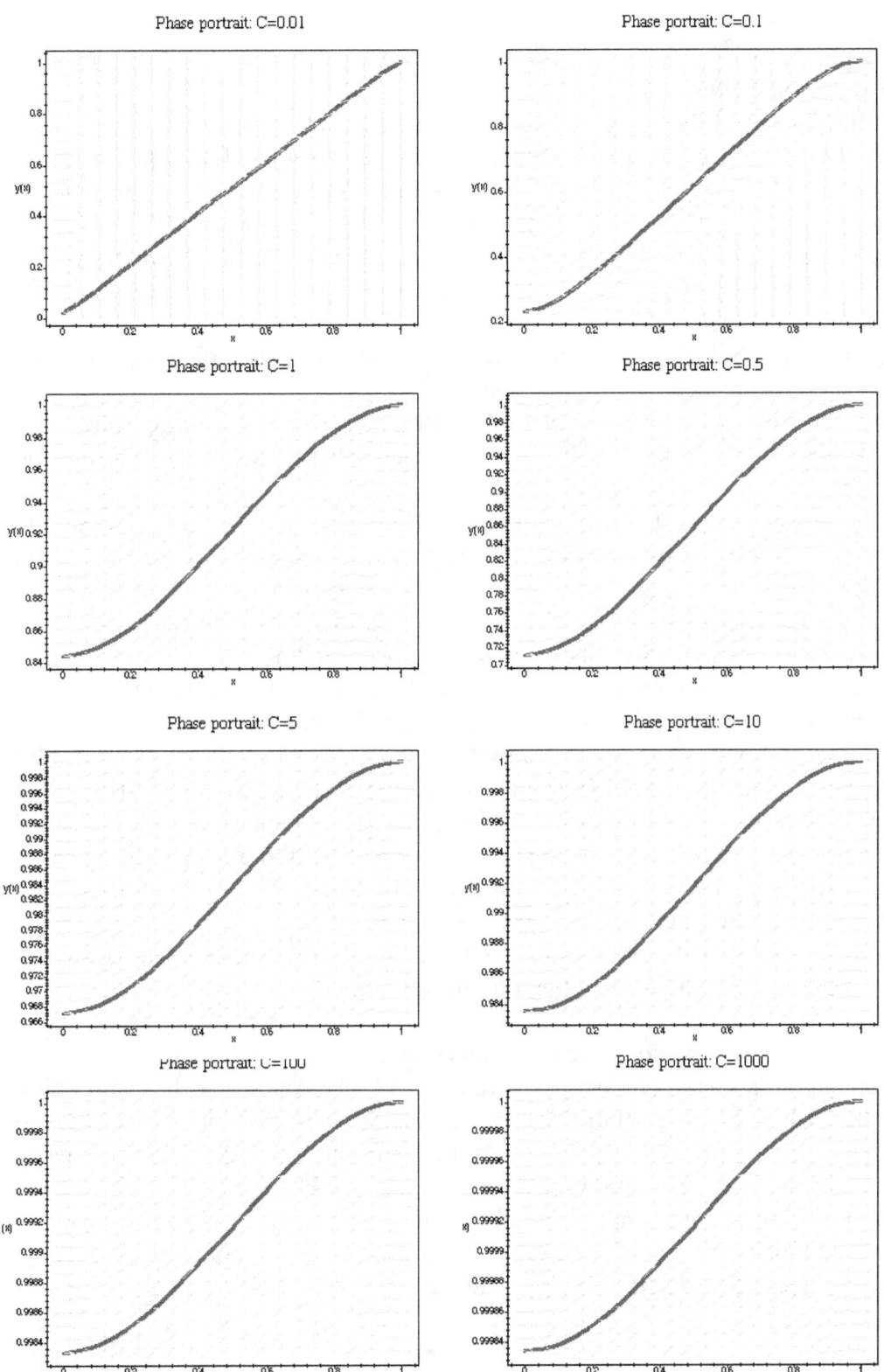

Figure B.2: Phase portraits of the ODE for various diffusion coefficients D.

phase portraits of the ODE for different values of the constant C, which is in major the diffusion coefficient.

x is the radius of the unreacted core and y is the radius of the overall particle. The difference $y - x$ is the film thickness and can be derived from the solid line in the portrait. For large values of D the film thickness is larger. The solid line represents the solutions of the ODE within the parameter space $0<x,y\leq 1$.

The explicit time dependence can not be derived directly from the portraits, but the film thickness as a function of the overall sphere radius y or unreacted core radius x can be derived directly from the solid line in the plots.

For larger values of C a larger film thickness is obtained, which is expected when the the diffusion coefficient $\mathcal{D}_e$ becomes larger.

From the phase portraits the evolution of film thickness on a reacting sphere can be derived by using the linear relation between sphere thickness y and reaction time t: $y(t) = R_0 - \beta t$.

An according representation of evolution of film thickness $d = y - x$, unreacted core thickness x and sphere radius y is displayed in Figure B.3. Data were taken from the phase portrait with $C = 1$ in Figure B.2. At time $t = 0$ the sphere radius is set 1, and the unreacted core radius is also 1. The film thickness therefore is 0 at time $t = 0$. The decrease of sphere radius is linear according to $y = R_0 - \beta t$. The unreacted core radius decreases unlike in the case of a flat sample with a turning point. After particular reaction time the unreacted core vanishes (stage 4 in Figure B.1), and only film material is left. The evolution of film thickness is simply derived from the two former relations by subtraction. Note that the film thickness also has a turning point at the same time with unreacted core thickness. When the unreacted core vanishes, the film thickness has a maximum value. With proceeding activation, the film thickness is on the extrapolated line for the overall sample radius (marked with circles).

But several different mathematical solutions for above ODE exist, and only one yields the exact expression for the film thickness.

A more sophisticated mathematical analysis is required to perform these investigations, which is beyond the scope of this work.

Using an Ansatz and introducing a parameter n in the Ansatz yields (i) a solution of the ODE and (ii) a relation between parameter n and variable x, which has to be true. If the relation holds, than the Ansatz is a solution of the ODE. It remains open whether further solutions of the ODE exist, which are not described by the Ansatz function. Additionally a relation (equation) between parameter n and the time t is found. This equation has several real and complex roots of $n(t)$. Finally, the film thickness is expressed by a function of $n(t)$.

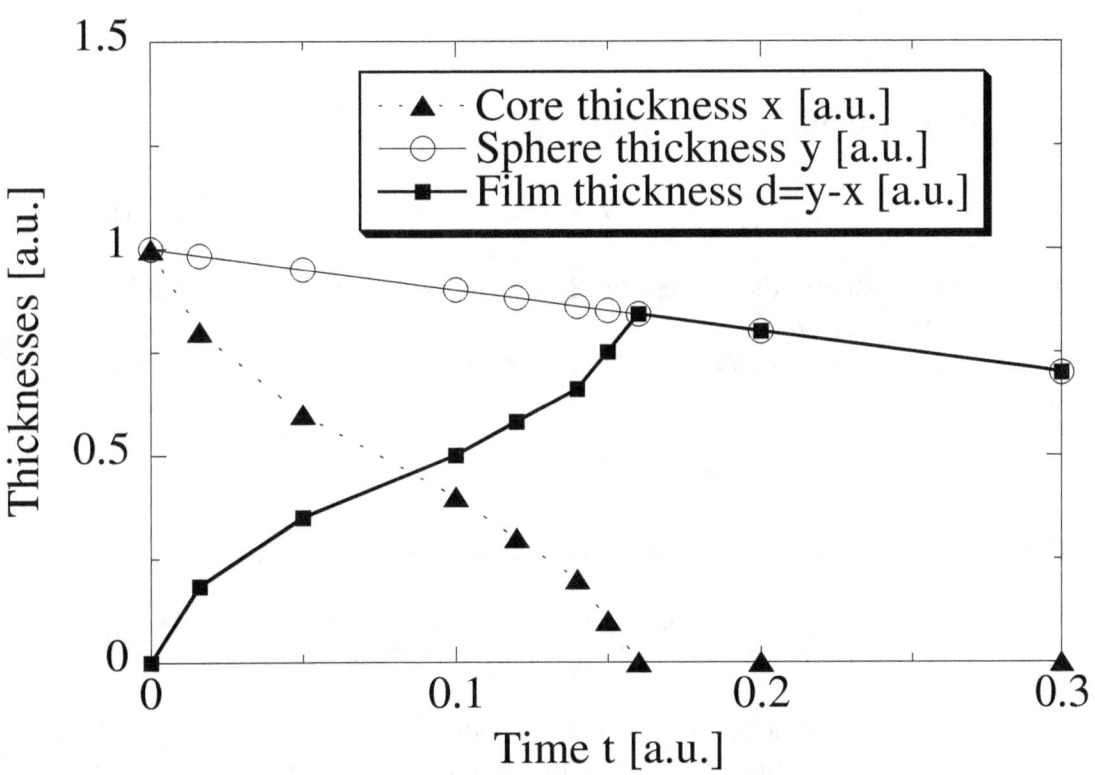

Figure B.3: Film thickness (filled square), sphere thickness (open circle) and unreacted core thickness (filled triangle) as a function of reaction time, as obtained from the phase portraits in Figure B.2. with C=1.

To solve the Abel ODE, we begin with following Ansatz:

$$y(x) = \exp\left(\frac{nx}{C}\right) \tag{B.16}$$

$$y'(x) = \frac{n}{C}y, \tag{B.17}$$

n being an arbitrary parameter: n ∈ R.
Inserting B.16 and B.17 in B.15 then yields the quadratic function

$$ny^2 - xy + x^2 = 0 \tag{B.18}$$

with the following two real valued roots,

$$y_\pm(x) = \frac{x}{2n}\left(1 \pm \sqrt{1-4n}\right), \tag{B.19}$$

provided the following constraint is true:

$$n \leq \frac{1}{4}. \tag{B.20}$$

With above constraint, the expression B.16 (Ansatz) is a real-valued solution for ODE in equation B.17.
Comparing equations B.16 and B.19, a relation between x and the number n is obtained:

$$y(n,x) = \exp\frac{nx}{C} = \frac{x}{2n}\left(1 \pm \sqrt{1-4n}\right) \tag{B.21}$$

The solutions of this equation were found using *Maple* and equals

$$x_\pm(n) = -\frac{C}{n} \cdot W\left(\frac{\pm 2n^2}{C\left(1 \pm \sqrt{1-4n}\right)}\right). \tag{B.22}$$

Equation B.22 determines a functional relationship between coordinate x and the parameter n. The argument $arg(W)$ of the LambertW function in equation B.22 will be termed $\mathcal{K}$ furtheron in this appendix; $\mathcal{K} = \mathcal{K}(n)$.
By some rearrangement we find

$$\frac{nx}{C} = -W(\mathcal{K}(n)) \tag{B.23}$$

and therefore, regarding above Ansatz in B.16:

$$y_\pm(n) = \exp(-W(\mathcal{K}(n))). \tag{B.24}$$

240 APPENDIX B. FILM GROWTH MODEL FOR SPHERICAL PARTICLES

We may remark that the overall sphere size at time t is

$$y(t) = R_0 - \beta t \tag{B.25}$$

By combining equations B.24 and B.25 we find the relationship between the time t and the parameter n.

$$\exp(-W(\mathcal{K}(n))) = R_0 - \beta t \tag{B.26}$$

or, by taking the logarithm on each side of equation B.26:

$$W(\mathcal{K}(n)) = -\ln(R_0 - \beta t). \tag{B.27}$$

We know y and x already as functions of the parameter n.
To proceed to the solution

$$d(t) = y - x = \ldots,$$

the relationship between the time t and the parameter n must be known.
As we need a function $n = n(t)$, we need the inverse of the LambertW function, which can be easily found by the definition for the W-function itself:

$$W(x) \cdot \exp(W(x)) = x \tag{B.28}$$
$$W^{-1}(x) = x \cdot \exp(-W(x)). \tag{B.29}$$

Using the argument of the W-function, we get

$$-\mathcal{K}(n) \cdot \exp(\mathcal{K}(n)) = \ln(R_0 - \beta t). \tag{B.30}$$

Using *Maple*, we find that equation B.30 has two complex and one real roots of the type $n = n(t)$.

$$n_{real}(t) = \frac{1}{4} - \frac{1}{4}\left(\frac{1}{3}B^{1/3} + \frac{4}{3}\frac{1}{B^{1/3}} - \frac{1}{3}\right)^2 \tag{B.31}$$

$$n_{imag}^{+}(t) = \frac{1}{4} - \frac{1}{4}\left(-\frac{1}{6}B^{1/3} - \frac{2}{3}\frac{1}{B^{1/3}} - \frac{1}{3} + \frac{1}{2}I\sqrt{3}\left(\frac{1}{3}B^{1/3} - \frac{4}{3}\frac{1}{B^{1/3}}\right)\right)^2 \tag{B.32}$$

$$n_{imag}^{-}(t) = \frac{1}{4} - \frac{1}{4}\left(-\frac{1}{6}B^{1/3} - \frac{2}{3}\frac{1}{B^{1/3}} - \frac{1}{3} - \frac{1}{2}I\sqrt{3}\left(\frac{1}{3}B^{1/3} - \frac{4}{3}\frac{1}{B^{1/3}}\right)\right)^2 \tag{B.33}$$

In equations B.31 - B.33 following abbreviations A and B were used:

$$A := \mathrm{LambertW}(-\ln(R - \beta t))\tag{B.34}$$

$$B := 8 + 108\,A\mathcal{C} + 12\,\sqrt{12\,A\mathcal{C} + 81\,A^2\mathcal{C}^2}\tag{B.35}$$

As in equation B.13 four different combinations of signs occur, also 4 different expressions for $x_\pm(n)$ are obtained.
Therefore 12 different functions $n(t)$ are obtained.
Combining y(n) in equation B.25 and n(t) in equation B.31, the function y=y(t) can be obtained directly. It remains to calculate the film thickness

$$d(t) = y - x.\tag{B.36}$$

It turns out that the following way is adequate for proceeding to the solution: For the W-function the general relationship holds [89] at $x \neq 0$:

$$W'(x) = \frac{1}{1+W} \cdot \exp(-W(x)) = \frac{W(x)}{x(1+W(x))}\tag{B.37}$$

Therefore following relation is valid:

$$W'(1+W) = \exp(-W(x)) = \frac{W(x)}{x}\tag{B.38}$$

From equations B.23 and B.24 we therefore find y as a function only explicit of n:

$$y(n) = \exp(\mathcal{K}) = \frac{W(\mathcal{K})}{\mathcal{K}}\tag{B.39}$$

Now x must be expressed as a function on n. From the Ansatz B.16 we find

$$\ln y = \frac{nx}{\mathcal{C}}.\tag{B.40}$$

Rearranging yields

$$\frac{\mathcal{C}}{n}\ln y = x.\tag{B.41}$$

Inserting equation B.24 in equation B.41 yields

$$\frac{\mathcal{C}}{n}\ln(\exp(-W(\mathcal{K}))) = x \to x(n) = -\frac{\mathcal{C}}{n}W(\mathcal{K}(n)).\tag{B.42}$$

Inserting equations B.39 and B.42 in equation B.36 yields

$$d(n) = y(n) - x(n) = \frac{W(\mathcal{K}(n))}{\mathcal{K}(n)} + \frac{C}{n}W(\mathcal{K}(n)) = \left(\frac{C}{n} + \frac{1}{\mathcal{K}(n)}\right) \cdot W(\mathcal{K}(n)),$$
(B.43)

$\mathcal{K}(n)$ being the argument of the LambertW function in equation B.22.
Equation B.43 is the film thickness as a function of the parameter n, which is a function of the reaction time t according to equations B.31 - B.33.
For each sign of the expressions $x_{\pm}(n)$ and $n_{\pm}(t)$ in equations B.22 and B.31 terms are built, therefore yielding 4 solutions for the real values for n and 8 solutions for the imaginary values for n.
The curves for the real values are plotted in Figure B.4 versus the activation time for different constants C (C = 0.1, 1, and 10).

The right choice of the curves for actual film thickness depends on initial and boundary conditions, which were not taken into account so far. There exist solutions for film thicknesses at $t < 0$, but they are physically meaningless. Note that the curvature is identical (turning point) with those displayed in Figure B.3, except the diverging film thickness at times where the unreacted core vanishes (stage 4 in Figure B.1) and the film thickness necessarily must decrease. For increasing constant C (diffusion coefficient), the film thickness after particular activation time also increases, which is expected by theory.
The film thicknesses obtained with the imaginary functions of $n(t)$ exhibit similar curvatures as those with real $n(t)$.
The thicknesses for times $t < 0$ are physically meaningless, but are plotted for completeness in understanding of the nature of the solutions. At time $t = 0$ the thicknesses equal 1, because initial and boundary conditions were not taken into account yet.
Note that the curvature of the thickness curves corresponds quite to the thickness behaviour displayed in Figure B.3., with the exception that a divergence is shown for the times, at which the unreacted core vanishes and the film thickness yields a maximum value.
With increasing diffusion coefficient (in C), the film thickness after a particular activation time increases, which is expected by theory.

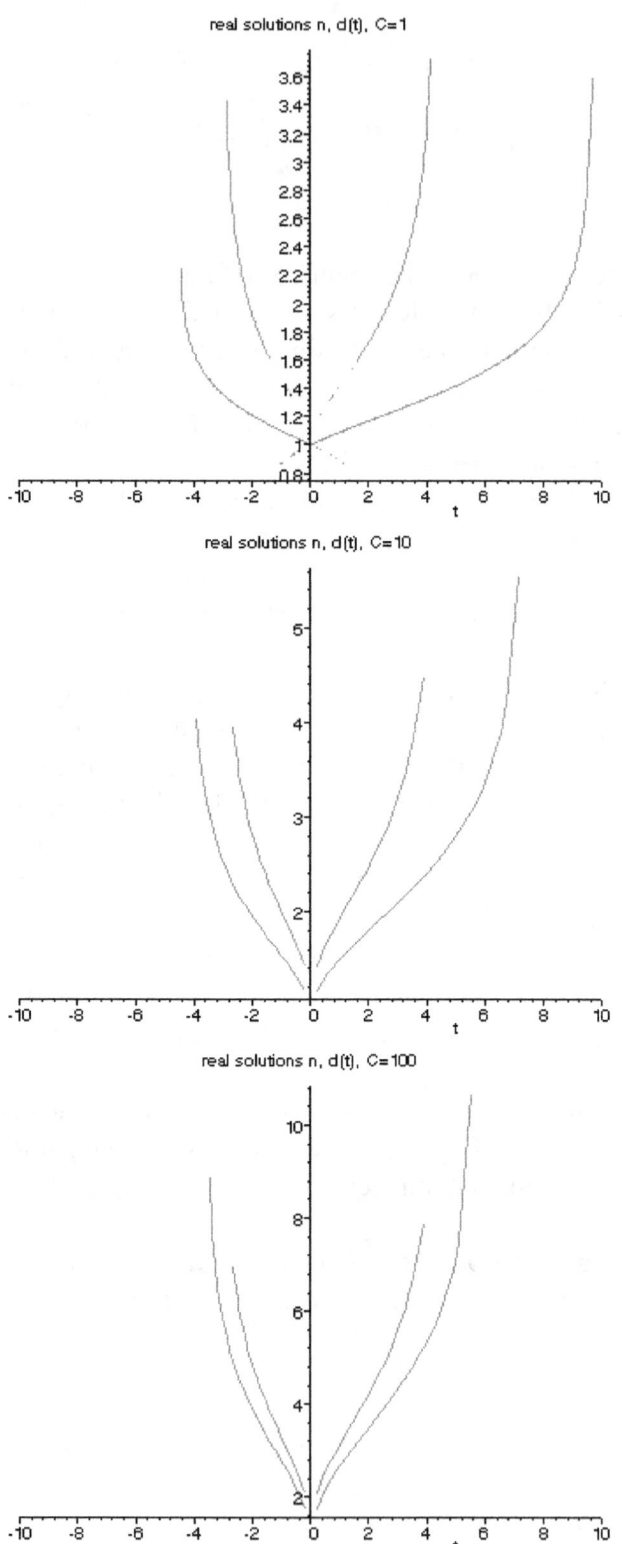

Figure B.4: Plots of film thickness versus reaction time, as calculated from equation B.43 for 4 different and real functions of $n(t)$, $R = 1$, and $\beta = 0.1$.

Bibliography

[1] Dr. W. Paul. private communication. 1996. ABB Corporate Research, Applied Physics Division.

[2] A. Braun, M. Bärtsch, B. Schnyder, R. Kötz, O. Haas, M. Carlen, T. Christen, C. Ohler, P. Unternährer, and E. Krause. A study on oxidised glassy carbon sheets for supercapacitor electrodes. In D. Doughty., editor, *New Materials for Batteries and Fuel Cells*, MRS Proceedings, Pasadena, 1999. MRS, Materials Research Society.

[3] R. Kötz, M. Bärtsch, A. Braun, and O. Haas. Bipolar electrochemical capacitor stacks with glassy carbon electrodes. In *The 7th International Seminar on Double Layer Capacitors and Similar Energy Storage Devices*, number 7, Deerfield Beach, Florida, Dec 1997.

[4] A.F. Burke and T.C. Murphy. Materials for electrochemical energy storage and conversion - batteries, capacitors and fuel cells. In D.H. Doughty, B. Yvas, T. Takamura, and J.R. Huff, editors, *Materials Research Society - Symposium Proceedings*, number 393, pages 375–395, 1995.

[5] Friedrich Hund. *Wissenschaftsgeschichtliche Anhänge*. Bibliographisches Institut VEB, Leipzig, Leipzig/Germany, 1951.

[6] Hamann and Vielstich, editor. *Elektrochemie I*. Number 41 in taschentext. Verlag Chemie, Physik Verlag, Weinheim, 1975. ISBN 3-527-21039-3,ISBN 3-87664-539-5.

[7] Hamann and Vielstich, editor. *Elektrochemie II*. Number 42 in taschentext. Verlag Chemie, Physik Verlag, Weinheim, 1975. ISBN 3-527-2108-,ISBN 3-87664-5398-5.

[8] Allen J. Bard and Larry R. Faulkner. *Electrochemical Methods - Fundamentals and Applications*. John Wiley & Sons Inc., New York, Chichester, Brisbane, Toronto, first edition, 1980.

[9] U. Stimming. Introduction to Electrochemistry. In W. Schilling, K. Urban, and H. Wenzl, editor, *Elektrokeramische Materialien - Grundlagen und Anwendungen*, number 26 in Vorlesungsmanuskripte des IFF-Ferienkurs, pages D3.1 – D3.44, J March 1995.

[10] L.D. Landau and E.M. Lifschitz. *Statistische Physik*. Number V in Lehrbuch der Theoretischen Physik. Akademie-Verlag Berlin, Leipziger Straße 3-4, Berlin, O-1086, Germany, 1987. ISBN 3-05-500063-3, Bd. V-ISBN 3-05-500069-2.

[11] N. Ibl. *Current Distribution*. Number 6. Plenum Publishing Corporation, 1982.

[12] M. Bärtsch, R. Kötz, A. Braun, and O. Haas. High-power electrochemical double-layer capacitor. In *Proceedings of the 38th Power Sources Conference*, number 5 in 1, pages 17–21, Boston, MA, Jun 1998.

[13] M. Sullivan, R. Kötz and O. Haas. In *Proceedings of the Symposium on Electrochemical Capacitors*, 95-29, pages 198–209. The Electrochemical Society, ECS, 1996.

[14] S. Sarangapani, B.V. Tilak, and C.P. Chen. *J. Electrochem. Soc.*, 143(11):3791 3799, 1996.

[15] M. Pourbaix. *Atlas of Electrochemical Equilibria in Aqueous Solutions*. International Tables for Crystallography. National Association of Corrosion Engineers, Houston, second edition, 1974. Section 17.1.

[16] G. M. Jenkins and K. Kawamura. *Polymeric carbons - carbon fibre, glass and char*. Cambridge University Press, Cambridge UK, first edition, 1976.

[17] I. Tanahashi and A. Yoshida and A. Nishino. *Carbon*, 29:1033, 1991.

[18] Electrochemical Industry Association. *Statistical Data*. Electrochemical Industry Association, Tokyo, Japan, May 1993.

[19] Electrochemical Industry Association. *Statistical Data*. Electrochemical Industry Association, Tokyo, Japan, May 1994.

[20] M.S. Dresselhaus and G. Dresselhaus and P. Eklund. *Science of Fullerenes and Carbon Nanotubes*. Academic Press, Jan 1996. ISBN 0-12-221820-5.

[21] F.C. Cowlard and J.C. Lewis. *J. Mater. Sci.*, 2:507–512, 1967.

[22] J.C. Lewis, B. Redfern, and F.C. Cowlard. *Solid-State Electronics*, 6:251–254, 1963.

[23] V.A. Drits and C. Tchoubar. *X-ray Diffraction by Disordered Lamellar Structures*. Springer Verlag, 1990.

[24] W.S. Rothwell. *J. Appl. Phys.*, 39(3):1840–1844, February 1968.

[25] R. Perret and W. Ruland. *J. Appl. Cryst.*, 1:308–313, 1968.

[26] F.C. Cowlard and J.C. Lewis. *J. Mater. Sci.*, 2:507–512, 1967.

[27] J. Miklos, K. Mund, and W. Naschwitz. Offenlegungsschrift de 30 11 701 a1. *Patent of Siemens AG*, Offenlegungsschrift DE 30 11 701 A, 1980.

[28] Z. Laušević and G.M. Jenkins. *Carbon*, 24(5):651–652, 1986.

[29] Max Born. *Optik - Ein Lehrbuch der elektromagnetischen Lichttheorie*. Springer Verlag, Stuttgart, 3. Auflage, 2. Nachdruck edition, 1985. 45, Seite 145.

[30] W. Xing, J. S. Xue, T. Zheng, A. Gibaud, and J. R. Dahn. *J. Electrochem. Soc.*, 143(11):3482–3491, Nov 1996.

[31] R.T. DeHoff, E.H. Aigeltinger, and K.R. Craig. Experimental determination of the topological properties of three dimensional microstructures. *J. Microsc.*, 95:69–91, 1972.

[32] W.R. Thompson. The geometric properties of microscopic configurations. I. General aspects of projectometry. *Biometrica*, 24:21–26, 1932.

[33] D.C. Sterio. The unbiased estimation of number and sizes of arbitrary particles using the disector. *J. Microsc. Journal*, 134:127–136, 1984.

[34] H.J. Gundersen, R.W. Boyce, J.R. Nyengaard, and A. Odgaard. The ConnEuler: unbiased estimation of connectivity using physical disectors under projection. *Bone*, 14:217–222, 1993.

[35] H.J. Vogel. Morphological Determination of pore connectivity as a function of pore size using serial sections. *Europ. J. Soil Sci.*, 48:365–377, Jun 1997.

[36] H.J. Vogel and A. Kretzschmar. Topological characterization of pore space in soil - sample preparation and digital image processing. *Geoderma*, 73:23–38, 1996.

[37] H.J. Vogel. Die Kontinuität des Porenraumes in Abhängigkeit der Porengrösse - eine morphologische Materialfunktion. *Mitt. Dt. Bodenkundl. Gesellsch.*, 80, 1996.

[38] Derrick Mancini. private communication. 1999. Argonne National Laboratory.

[39] J. Miklos, K. Mund, and W. Naschwitz. Siemens AG, Offenlegungsschrift DE 30 11 701 A1, Deutsches Patentamt, 1980.

[40] R.C. Bansal, J.B. Donnet, and H.F. Stoeckli. *Active Carbon*. Marcel Dekker, New York - Basel, 1988.

[41] Y. Liu, J.S. Xue, T. Zheng, and J.R. Dahn. *Carbon*, 34(4):499–503, 1996.

[42] W. Xing, J. S. Xue, and J. R. Dahn. *J. Electrochem. Soc.*, 143(10):3046–3052, oct 1996.

[43] M. Sullivan, M. Bärtsch, R. Kötz and O. Haas. In *Proceedings of the Symposium on Electrochemical Capacitors*, 96-25, pages 192–201. The Electrochemical Society, ECS, 1997.

[44] *Alphabetical Index of Inorganic Phases*. International Tables for Crystallography. International Center for Diffraction Data (ICDD), 1601 Park Lane, Swarthmore, Pennsylvania 19081-2389, USA, 1992.

[45] A. Braun, R. Kötz, and R. Saliger. Saxs on oxidised glassy carbon sheets. In W. Laasch, G. Materlik, J.R. Schneider, and H. Schulte-Schrepping., editors, *HASYLAB Annual Report 1998*, number 2263 in Contributions Part I, Notkestraße 23, D-22603 Hamburg, Germany., publisher =, 1998.

[46] H.-G. Haubold et. al. *Rev. Sci. Instrum*, 60:1943–1946, 1989.

[47] A.J.C. Wilson. *Mathematical, Physical and Chemical Tables*, volume C of *International Tables for Crystallography*. Kluwer Academic Publishers, Dordrecht, first edition, 1992.

[48] A. Gupta and I.R. Harrison and J. Lahijani. *J. Appl. Cryst*, 27:627–636, 1994.

[49] S. Brunauer and P.H. Emmett and E. Teller. *J. Amer. Chem. Soc.*, 60:309, 1938.

[50] P.A. Webb and C. Orr. *Analytical Methods in Fine Particle Technology*. Micromeritics Instrumental Corporation, 1997. ISBN 0-9656783-0-X.

[51] S.J. Gregg and K.S.W. Sing. *Adsorption Surface Area and Porosity*. Academic Press, London, 2 edition, 1982.

[52] Kenneth S.W. Sing. The Use of Physisorption for the Characterization of Microporous Carbons. *Carbon*, 27(1):5–11, 1989.

[53] K.S.W. Sing, D.H Everett, R.A.W. Haul, L. Moscou, R.A. Pierotti, J. Rouquerol, and T. Siemieniewska. *Pure Appl. Chem.*, 57:603, 1985.

[54] L.D. Landau. The Scattering of X-rays by Crystals with Variable Lamellar Structure. $\Psi \Xi\ T\Phi$, 12:1227, 1937.

[55] B.E. Warren and N.S. Gingrich. *Phys. Rev.*, 46, 1934.

[56] H. Shi and J. Barker and M. Y. Saïdi and R. Koksbang. *J. Electrochem. Soc.*, 143(11):3466–3472, November 1996.

[57] V.A. Drits and C. Tchoubar. *X-ray Diffraction by Disordered Lamellar Structures*. Springer Verlag, 1990.

[58] M.A. Short and P.L. Walker. *Carbon*, 1(3), 1963.

[59] N.N. Greenwood and A. Earnshow. *Chemistry of the Elements*. Pergammon Press, New York, 1984. p. 296.

[60] IUPAC. Reporting Physisorption Data for Gas/Solid Systems. *Pure & Appl. Chem.*, 57(4):603–619, 1985. Pores with widths exceeding about 50 nm are called macropores, pores of widths between 2 nm and 50 nm are called mesopores, and pores with widths not exceeding about 2 nm are called micropores.

[61] A. Guinier and G. Fournet. *Small Angle Scattering of X-Rays*. John Wiley & Sons, Inc., New York, 1955.

[62] A. Guinier. *Ann. Phys.*, 12:161, 1939.

[63] A. Guinier. *Theorie et technique de la radiocristallographie*. Dunod, Paris, 1964.

[64] V. Gerold. *Zeitschr. f. Angew. Phys.*, IX:43–44, 1957.

[65] A. Braun, M. Bärtsch, O. Merlo, B. Schnyder, R. Kötz, and O. Haas. Thermally activated glassy carbon - a material for supercapacitor electrodes. *PSI Annual Report 1997*, V:46–47, 1998. Annex V, General Energy Research.

[66] H.-G. Haubold. Röntgenkleinwinkelstreuung an synchrotronstrahlungsquellen. In *Synchrotronstrahlung zur Erforschung kondensierter Materie.*, number 23 in Vorlesungsmanuskripte des IFF-Ferienkurses, page 29.1, Jülich, March-April 1992. Forschungszentrum Jülich GmbH, Institut f. Festkörperforschung. ISBN 3-89336-088-3.

[67] O. Glatter and O. Kratky. *Small Angle X-ray Scattering*. Academic Press, London, 1982.

[68] G. Porod. *Kolloid Z.*, 124(2):83–114, 1951.

[69] G. Porod. *Kolloid Z.*, 125(2):51–122, 1951.

[70] N.A. Seaton, J.P.R.B. Walton, and N. Quirke. A New Analysis Method for the Determination of the Pore Size Distribution of Porous Carbons from Nitrogen Adsorption Measurements. *Carbon*, 27(6):853–861, 1989. The lognormal distributiuon has the advantage that the product of f(w) and a power of w is analytically integrable, which in turn means that eqn (2) can be evaluated analytically. To check that the fit is not unduly constrained by the lognormal form f(w), the bimodal gamma distribution was selected as an alternative functional form. Pore size distributions obtained usiong the two functional forms were compared for several samples and found to be almost identical. Finally, it must be emphasised that the ultimate justification for the use of any functional form of f(w) is its ability to fit experimental isotherms.

[71] Lord Rayleigh. *Proc. Roy. Soc.(London)*, A-84(25), February 1911.

[72] A. Braun, M. Bärtsch, B. Schnyder, R. Kötz, and O. Haas. Saxs on glassy carbon with variable scattering contrast. In Albert Furrer, editor, *Complementarity between Neutron and Synchrotron X-ray Scattering.*, Proceedings of the Sixth Summer School on Neutron Scattering., page 366, Singapore, 1998. World Scientific.

[73] H.-G. Haubold. private communication. 1998. Forschungszentrum Jülich.

[74] W. Ruland. *J. Appl. Cryst.*, 7:383–386, 1974.

[75] W. Ruland. *J. Appl. Cryst.*, 4:70–73, 1971.

[76] H. Peterlik, P. Fratzl, and K. Kromp. *Carbon*, 32(5):939, 1994.

[77] S.K. Sinha. Small-angle and surface scattering from porous and fractal materials. In Albert Furrer, editor, *Complementarity between Neutron and*

Synchrotron X-ray Scattering., Proceedings of the Sixth Summer School on Neutron Scattering., Singapore, 1998. World Scientific.

[78] D. Laser and M.J. Ariel. *J. Electroanal. Chem.*, 80:291, 1974.

[79] H. Gunasingham and B. Fleet. *Analyst*, 107:896, 1982.

[80] B.D. Epstein, E. Dalle-Malle, and J.S. Mattson. *Carbon*, 9:609, 1971.

[81] F. Blurton. *Electrochim. Acta*, 18:869, 1973.

[82] A. Procter and P.M.A. Sherwood. *Carbon*, 21:53, 1983.

[83] J.F. Evans and T. Kuwana. *Anal. Chem.*, 49:1635, 1977.

[84] L.G.J. de Haart. Impedanzspektroskopische Untersuchungen an elektrokeramischen Materialien. In W. Schilling, K. Urban, and H. Wenzl, editor, *Elektrokeramische Materialien - Grundlagen und Anwendungen*, number 26 in Vorlesungsmanuskripte des IFF-Ferienkurs, pages B1 – B14, J March 1995.

[85] F.M. Delnick. Carbon Supercapacitors. In *Symposium on Science of Advanced Batteries*, Cleveland, Ohio, nov 1993. U.S. Dept. of Energy, DOE. Cleveland, Ohio.

[86] R. de Levie. *Electrochemistry*, volume 6 of *Advances in Electrochemistry and Electrochemical Engineering*, pages 308–313. Interscience Publishers, Wiley & Sons, New Yourk, London, Sydney, 1967.

[87] M. Bärtsch. private communication.

[88] J.C. Panitz. private communication. 1998. PSI.

[89] K.M. Briggs. W-ology or some exactly solvable growth models. *to be published*, 1998. private communication, and http://www.epidem.plantsci.cam.ac .uk/ kbriggs/W-ology.html Dept. Plant Sciences, Univ. of Cambridge, UK.

[90] A. Braun, M. Bärtsch, B. Schnyder, and R. Kötz. A model for the film growth in samples with two moving reaction frontiers - an application and extension of the unreacted-core model. *Chem. Eng. Sci.*, 55:5273–5282, 2000.

[91] R.M. Corless, G.H. Gonnet, D.E.G. Hare, D.J. Jeffrey, and D.E. Knuth. On the Lambert W Function. *Maple Share Library*, 1991.

[92] F.N. Fritsch, R.E Shafer, and W.P. Crowley. *Comm. ACM*, 16:123–124, 1973. Algorithm 443, Solution of the Transcendental Equation: w*(exp (w)) = x.

[93] Octave Levenspiel. *Chemical Reaction Engineering*. John Wiley & Sons, NY, second edition, 1972.

[94] M. Ishida and C.Y. Wen. Comparison of kinetic and diffusional models for solid-gas reactions. *AIChE Journal*, 14:311–317, 1968.

[95] M. Ishida and C.Y. Wen. *Chem. Eng. Sci.*, 26:1031–1041, 1971.

[96] M. Ishida, C.Y. Wen, and T. Shirai. *Chem. Eng. Sci.*, 26:1043–1048, 1971.

[97] E. Comparini and R. Ricci. Convergence to the pseudo-steady-state approximation for the unreacted core model. *Appl. Anal.*, 26(4):303–325, 1988.

[98] A. Di Liddo and I. Stakgold. *J. Math. Anal. Appl.*, 152:584–599, 1990.

[99] R.E. Carter. Kinetic Model for Solid-State Reactions. *J. Chem. Phys.*, 34(6):2010–2015, jun 1961. The parabolic law has been repeatedly shown to be valid for plane films and is so well accepted that experimental agreement with eqn (9) (dy/dt=k/y) has become the criterion for determining if an oxidation reaction is diffusion controlled.

[100] J. Szekely, J.W. Evans, and H.Y. Sohn. *Gas-Solid Reactions*. Academic Press, NY, 1976. ISBN 0-12-680850-3.

[101] A. Rehmat, S.C Saxena, R. Land, and A.A. Jonke. *Can. J. Chem. Eng.*, 56(25):316–322, 1978.

[102] S.V. Sotirchos and N.R. Amundson. Dynamic Behavior of a Porous Char Particle Burning in an Oxygen-Containing Environment.

[103] A. Braun, M. Bärtsch, B. Schnyder, R. Kötz, O. Haas, G. Goerigk, and H.-G. Haubold. X-ray scattering and adsorption studies of thermally oxidized glassy carbon. *J. Non-Cryst. Sol.*, 260:1–14, 1999.

[104] C. Wagner. *Z. Phys. Chem. B*, 21:25, 1933.

[105] G. Tammann. *Z. Anorg. Allg. Chemie* , 111:78–89, 1920.

[106] B.W. Char and K.O. Geddes and G.H. Gonnet and B.L. Leong and M.B. Monagan and S.M. Watt. *The Maple V Language Reference Manual*. Springer Verlag, 1991.

[107] C. Wagner. *Z. Phys. Chem. B*, 32:417, 1936.

[108] C. Wagner. *Z. Phys. Chem. B*, 40:455, 1938.

[109] R.E. Carter. Addendum: Kinetic model for solid-state reactions. *J. Chem. Phys.*, 34(6):1137–1138, 1961.

[110] H.J. Siebeneck, P.A. Urick, D.P.H. Hasselman, E.J. Minford, and R.C. Bradt. *Carbon*, 15:187–188, 1977.

[111] O. Merlo. Einfluss des Aktivierungsmediums und der Aktivierungszeit auf die Kapazität von Glassy Carbon. PSI, September 1997.

[112] A. Braun, M. Bärtsch O. Merlo, B. Schaffner, B. Schnyder, R. Kötz, O. Haas, and A. Wokaun. Evolution of electrochemical double layer capacitance in glassy carbon during thermal oxidation. *Carbon*. in press.

[113] V.I. Arnol'd. *Ordinary Differential Equations*. Springer-Textbook. Springer-Verlag, Berlin-Heidelberg, 1992. ISBN 3-540-54813-0,ISBN 0-387-54813-0.

[114] A. Braun, D. Alliata, M. Bärtsch, B. Schnyder, and R. Kötz. Sans on oxidised glassy carbon sheets. In Werner Wagner, editor, *SINQ Annual Report 1998-1999*, number NN in Contributions Part NN, Villigen PSI, 1999. SINQ.

[115] K. Kinoshita. *Carbon - Electrochemical and Physicochemical Properties*. John Wiley & Sons, 1988.

[116] H.E. Darling. *J. Chem. Eng. Data*, 9:421, 1964.

[117] V.M.M. Lobo. *Handbook of Electrolyte Solutions*, volume 41 of *Physical Sciences Data*. Elsevier. Part A.

[118] Joachim Grehn. *Metzler Physik*. J.B. Metzlersche Verlagsbuchhandlung, Stuttgart, Second edition, 1988.

[119] R. Gallay. Fonctionnement et technologie des condensateurs électrochimiques. Journée scientifique BOOSTCAP, Conference at montena condis, June 10, 1999.

[120] D. Alliata, P. Häring, O. Haas, R. Kötz, and H. Siegenthaler. In situ atomic force microscopy of electrochemically activated glassy carbon. *Electrochemical and Solid-State Letters*, 2(1):33–35, 1999.

[121] H. Maleki, L.R Holland G.M. Jenkins, R.L. Zimmerman, and W. Porter. *J. Mater. Res.*, 11:2368, 1996. Maximum heating rates for producing undistorted glassy carbon ware determined by wedge-shaped samples.

[122] R. E. Franklin. *Acta Cryst.*, 4:253–261, 1951.

[123] Y.A. Levendis and R.C. Flagan. *Carbon*, 27(2):265–283, June 1989.

[124] A. Braun, M. Bärtsch, R. Kötz, B. Schnyder, and O. Haas. Investigation of the porous structure of glassy carbon by saxs - an application of synchrotron radiation. *PSI Scientific Report 1998*, V:34, March 1999. ISSN 1423-7342.

[125] R. Saliger. private communication. 1997. Universität Würzburg.

[126] D.F.R. Mildner, R. Rezvani, P.L. Hall, and R.L. Borst. *Appl. Phys. Lett.*, 2:253, 1986.

[127] J.E. Martin. *J. Appl. Cryst.*, 19:25–27, 1986.

[128] M. Daoud and J.E. Martin. *Fractal Properties of Polymers*.

[129] H.D. Bale and P.W. Schmidt. *Phys. Rev. Lett.*, 53:596, 1984.

[130] P.W. Schmidt and H. Kaiser and A. Hohr and J.S. Lin and H.B. Neumann and D. Avnir. *J. Chem. Phys.*, 90:5016, 1989.

[131] P.Z. Wong and A. Bray. *J. Appl. Cryst.*, 21:786, 1988.

[132] P.Z. Wong and A. Bray. *Phys. Rev. B*, 37:7751, 1988.

[133] A.J. Hurd. private communication. 1999. Sandia National Laboratories.

[134] J. Koresh and A. Soffer. *J. Electrochem. Soc.*, 124:1379, 1977.

[135] I. Tanahashi, A. Yoshida, and A. Nishino. *Carbon*, 29:1033, 1991.

[136] A. Braun, M. Bärtsch, B. Schnyder, R. Kötz, O. Haas, and A. Wokaun. Evolution of bet internal surface area in glassy carbon powder during thermal oxidation. *Carbon*, 40(3):375–382, 2002.

[137] B. Schnyder, A. Braun, M. Bärtsch, and R. Kötz. to be published.

[138] W. Jander. *Z. anorg. u. allgem. Chem.*, 163(1), 1927.

[139] E. Kamke. *Differentialgleichungen: Lösungsmethoden und Lösungen*, volume 1. Akademische Verlagsgesellschaft, Leipzig, 1962. Gewöhnliche Differentialgleichungen.

www.ingramcontent.com/pod-product-compliance
Lightning Source LLC
Chambersburg PA
CBHW080955170526
45158CB00010B/2814